Soft Computing Techniques for Engineering Optimization

Science, Technology, and Management Series

Series Editor:
J. Paulo Davim

This book series focuses on special volumes from conferences, workshops, and symposiums, as well as volumes on topics of current interest in all aspects of science, technology and management. The series will discuss topics such as mathematics, chemistry, physics, materials science, nanosciences, sustainability science, computational sciences, mechanical engineering, industrial engineering, manufacturing engineering, mechatronics engineering, electrical engineering, systems engineering, biomedical engineering, management sciences, economical science, human resource management, social sciences and engineering education. This book series will present principles, models techniques, methodologies, and applications of science, technology and management.

Handbook of IOT and Big Data
Edited by Vijender Kumar Solanki, Vicente García Díaz, J. Paulo Davim

Advanced Mathematical Techniques in Engineering Sciences
Edited by Mangey Ram and J. Paulo Davim

Soft Computing Techniques for Engineering Optimization

Kaushik Kumar, Supriyo Roy, and
J. Paulo Davim

CRC Press is an imprint of the
Taylor & Francis Group, an **informa** business

CRC Press
Taylor & Francis Group
6000 Broken Sound Parkway NW, Suite 300
Boca Raton, FL 33487-2742

First issued in paperback 2021

ISBN 13: 978-0-367-78021-0 (pbk)
ISBN 13: 978-0-367-14861-4 (hbk)

Library of Congress Cataloging-in-Publication Data

Names: Kumar, K. (Kaushik), 1968- author. | Roy, Supriyo, author. | Davim, J. Paulo, author.
Title: Soft computing techniques for engineering optimization / authored by Kaushik Kumar, Supriyo Roy, and J. Paulo Davim.
Description: Boca Raton : Taylor & Francis, 2019. | Series: Science, technology, and management series
Identifiers: LCCN 2018050968| ISBN 9780367148614 (hardback : alk. paper) | ISBN 9780429053641 (e-book)
Subjects: LCSH: Soft computing. | Industrial engineering. | Mathematical optimization.
Classification: LCC QA76.9.S63 K85 2019 | DDC 006.3—dc23
LC record available at https://lccn.loc.gov/2018050968

Visit the Taylor & Francis Web site at
http://www.taylorandfrancis.com

and the CRC Press Web site at
http://www.crcpress.com

Contents

Section III Hands-On Training on Various Software Dedicated for the Usage of Techniques

Preface

The term 'optimize' is basically a way of journey 'to make perfect'. The word 'optimus'– 'the best'–was actually derived from *Opis*, a Roman goddess of abundance and fertility, and is said to be the wife of Saturn. By her, the Gods designated the earth, because the earth distributes all goods, i.e., riches, goods, abundance, gifts, munificence, plenty, etc.

Let us consider the design of a steel structure, where some of the members are described by ten design variables. Each design variable represents a number of universal beam sections from a catalogue of ten available sections. Assuming that one full structural analysis of each design takes 1 s on a computer, how much time would it take to check all the combinations of cross sections to guarantee an optimum solution? The result is 317 years (10^{10} s), and hence is the requirement of 'optimization'.

Optimization can be used in any field as it involves formulating process or products in various forms. It is the 'process of finding the best way of using the existing resources while taking into account all the factors that influence decisions in any experiment'. The final product meets the requirements not only from the availability but also from the practical mass production criteria.

This book is mainly aimed at two major objectives. First, it explains the basic concept of soft computing and its application towards optimization. Second, it provides hand-in-hand exposure by guiding the readers on the journey towards optimization via an implicational area of concern.

The main objective of this book is bidirectional: to develop soft computing and to explore its efficacy and application towards optimization. Its main emphasis is directed towards industrial engineering outlook. The target audience is academic students, researchers and industry practitioners, engineers, research scientists and academicians working in interdisciplinary fields.

The chapters are categorized into three sections: Section 1 – Introduction to Soft Computing Techniques, Section 2 – Various case studies comprising of industrial problems and their optimal solutions using different techniques and Section 3 – Hands-on training on various software packages dedicated for the usage of techniques.

Section 1 contains Chapters 1 and 2, Section 2 has Chapters 3–10 and Section 3 enlists Chapters 11–13.

Section 1 starts with *Chapter 1* introducingthe reader to the concept of optimisation and single objective mathematical programming technique. It continues with generalized reduced gradient and derivative-based optimization. Emphasis has been given on various traditional methods such as Descent Method, Gradient-Based Method, Steepest Descent Method and

Newton's Method. The chapter starts discussing about Soft Computing and its difference from Hard Computing. It ends with an elaboration of the relevance of soft computing towards optimal solution.

Chapter 2 is a description of various soft computing techniques with an elaboration on the history of heuristics or search algorithms. Various soft computing approaches under derivative-free optimization such as Genetic Algorithms, Simulated Annealing, Random Search Method and Downhill Simplex Method have been discussed elaborately. Techniques under stochastic optimization such as Particle Swarm Optimization and Ant Colony Optimization have also been covered in this chapter. The chapter ends with an elaborative discussion on the latest development of heuristics named after hybridization.

Section 2 contains *seven* chapters in which different industrial-oriented problems have been optimized using different techniques. *Chapter 3* takes up the optimization of input process parameters of Electrical Discharge Machining using Response Surface Methodology. The chapter deals with the impact of machining parameters and machining characteristics of EN 19 tool steel workpiece in electrical discharge machining. Input variables in machining taken are pulse off time, voltage and pulse current, whereas the material removal rate is the response that needs to be maximized. To develop empirical models and relate them with process variables along with its interactions with concerned response function, response surface methodology is incorporated. A confirmatory test was performed at the end to validate the chosen input parameters.

Chapter 4 aims to determine the effect of machining parameters in electrical discharge machining on the machining characteristics of EN 19 tool steel work sample. Material removal rate and five surface roughness parameters are considered as responses, while the machining variables are pulse current, pulse off time, voltage and pulse on time. Solving multi-variable and multi-objective functions is quite complex, and hence in such cases, it becomes necessary to change multi-objectives to a single performance index, which represents the overall quality index for multiple quality characteristics. Considering complexities, a Weighted Principal Component Analysis has been chosen for the same. Similar to an earlier problem, confirmatory test was performed at the end to evaluate the extent of optimization.

Chapter 5 utilizes the Taguchi Technique towards optimization of a single objective situation. Here, Turning Operation in computer numerical control (CNC) Lathe has been chosen with an aim to maximize the material removal rate. From different parameters that can be thought of as manufacturing objectives, material removal rate is examined as a factor which specifically influences machining hour rate and cost of machining. Machining parameters, in particular, depth of cut, feed rate and cutting rate, were taken into consideration. To evaluate the improvement increments in the signal-to-noise ratio from initial process parameters, optimal process parameters were evaluated, indicating authenticity of the technique and also its validation.

As the Taguchi Technique has a different set of equations for maximization and minimization, *Chapter 6* is written on minimization problem. The problem chosen here is minimization of overcut in Electrical Chemical Machining. Feed rate, Inter-electrode gap, electrolyte concentration, and voltage are taken as process parameters and overcut as an output response. Since overcut is to be made minimum for better accuracy of the process, minimization has been used.

It usually happens in some of the machine components where more than one output parameter, such as material removal rate and surface roughness, are equally important. In this sequel, *Chapter 7* considers Grey-Taguchi Technique, a multi-objective optimization tool towards better wear properties for Kaolin-reinforced polymeric composite. Here, three independent parameters, namely filler content, speed and normal load, were considered, and two output parameters, such as wear and coefficient of friction, were optimized simultaneously. A confirmatory test at the end of the chapter confirms the validity of the technique.

Chapter 8 aims at utilizing Fuzzy Logic Technique in the optimization of an industrial problem. Here, wear characteristics of silica gel-reinforced aluminium metal matrix composites are studied. To minimize the number of experiments, Taguchi strategy is used for deciding the design of experiments. This chapter discusses how fuzzy reasoning could be used to change multiple responses into a solitary response by utilizing multi-response performance index. Wear characteristics have been evaluated at different volume fractions of ceramic tested at dry sliding condition with different normal load and sliding speed.

In *Chapter 9,* multi-objective optimization has been performed using Non-Dominated Sorting Genetic Algorithm. The problem chosen is non-conventional machining of an engineering material in electro-chemical machine. Material removal rate and surface roughness, of opposing nature, are an exceptional mix of machining parameters, which give better surface finish, and also high material removal rate was identified at the same time using the Non-Dominated Sorting Genetic Algorithm.

Chapter 10, the last chapter of Section 2, illustrates the usage of Artificial Bee Colony algorithm for optimization of surface roughness in Wire Electro-Discharge Machine. The aim of the chapter is to determine the effects of chosen process parameters, i.e., current, pulse on time and pulse off time and voltage, towards minimum surface roughness for EN 31 tool steel. As a validation test, the surface morphology was studied with the aid of scanning electron microscopy at optimal and average process parametric values.

Section 3, comprising three chapters, starts from this junction. *Chapter 11* provides a hands-on training on Minitab software, the most sought out form of optimization solution. Here, the performance of a chemical reaction under various controlled factors, such as temperature, pressure, concentration and surface area, is investigated by optimizing heat release.

Chapter 12 explores MATLAB® by taking the example used in Chapter 11 but performs multi-objective optimization using Artificial Neural Network. Here, the performance of a chemical reaction under various controlled factors, such as temperature, pressure, concentration and surface area, was optimized considering heat and CO_2 release as output. Readers are provided with snapshots of the screen output after every command, providing them with a user-friendly training programme.

Chapter 13, the last chapter of Section 3 and the book, is identical to the earlier chapter as far as the software utilized is concerned, but uses Genetic Algorithm for the process of optimization. Multi-objective optimization with two objective functions were discussed and depicted.

First and foremost, we thank God. In the process of putting this book together, we realized how true this gift of writing is for us. God has given us the power to believe in our passion and pursue our dreams. We could never have done this without the faith we have in God, 'The Almighty'.

We also thank our colleagues, friends and students of different highest level learning institutes and organizations. This book was not only inspired by them but also directly improved by their active involvement in its development. We look forward to discussing this book with them at future gatherings, as we are sure they will all need and read this soon.

We owe a huge thanks to all of our technical reviewers, editorial advisory board members, book development editor and the team of CRC Press personnel for their availability to work on this huge project. All their efforts have helped to make this book complete, and we couldn't have done it without their constant coordination and support.

Throughout the process of writing this book, many individuals, from different walks of life, have taken time to help us out. Last, but definitely not least, we thank our well-wishers for providing us encouragement and kind-hearted support. We would have probably given up without their support.

Kaushik Kumar
Supriyo Roy
J. Paulo Davim

please contact:
The MathWorks, Inc.
3 Apple Hill Drive
Natick, MA 01760-2098 USA
Tel: 508-647-7000
Fax: 508-647-7001
E-mail: info@mathworks.com
Web: www.mathworks.com

Authors

Dr. Kaushik Kumar, BTech (mechanical engineering, REC (Now NIT), Warangal), MBA (marketing, IGNOU) and PhD (engineering, Jadavpur University), is presently an associate professor in the department of mechanical engineering, Birla Institute of Technology, Mesra, Ranchi, India. He has 15 years of teaching and research experience and more than 11 years of industrial experience in a manufacturing unit of global repute. His areas of teaching and research interests are quality management systems, optimization, non-conventional machining, computer aided design/computer aided manufacturing (CAD/CAM), rapid prototyping and composites. He has nine patents, 23 books, 12 edited books, 38 book chapters, 130 international journal publications, 18 international and eight national conference publications to his credit. He is on the editorial board and review panel of seven international and one national journal of repute. He has been felicitated with many awards and honours.

Dr. Supriyo Roy, MSc (statistics), MTech (ORI&BM), PhD (business model optimization), has been working as an associate professor at Birla Institute of Technology, Mesra, Ranchi. A postdoctoral fellow from Airbus Group Endowed (formerly EADS-SMI) Center for Sourcing and Management, IIM, Bangalore, Roy has more than 5 years of industrial experience and 15 years of experience in postgraduate teaching. His teaching interest comprises interdisciplinary areas of management science, production and operations management, decision science, supply chain management, soft computing for manufacturing, etc. He has to his credit published one book, ten international book chapters, 12 conference papers and more than 45 research papers in leading international as well as national journals. Dr. Roy is associated with many international and national journals of repute as an impanelled reviewer. He has been felicitated with an award of 'Outstanding Contribution to Teaching' in management science.

Prof. J. Paulo Davim received his PhD in mechanical engineering in 1997, MSc in mechanical engineering (materials and manufacturing processes) in 1991, licentiate degree (5 years) in mechanical engineering in 1986 from the University of Porto (FEUP); the aggregate title from the University of Coimbra in 2005 and DSc from London Metropolitan University in 2013. He is European Engineer FEANI (EUR ING) by fédération européenne d'associations nationales d'ingénieurs (FEANI) and senior chartered engineer at the Portuguese Institution of Engineers with MBA and specialist title in engineering and industrial management. Currently, he is a professor in the department of mechanical engineering of the University of Aveiro. He has more than 30 years of teaching and research experience in manufacturing, materials and mechanical engineering, with special emphasis on machining and tribology. Recently, he also has interest in management/industrial engineering and higher education for sustainability/engineering education. He has received several scientific awards. He has worked as an evaluator of projects for international research agencies as well as an examiner of PhD thesis for many universities. He is the editor in chief of several international journals, guest editor of journals, editor of books, series editor of books and scientific advisor for many international journals and conferences. At present, he is an editorial board member of 25 international journals and acts as a reviewer for more than 80 prestigious Web of Science journals. In addition, he has been an editor (and co-editor) of more than 100 books and an author (and co-author) of more than ten books, 80 book chapters and 400 articles in journals and conferences (more than 200 articles in journals indexed in Web of Science/h-index 45+ and SCOPUS/h-index 52+).

Section I

Introduction to Soft Computing Techniques

1

Introduction to Optimization and Relevance of Soft Computing towards Optimal Solution

1.1 Optimization

Optimization is a systematic procedure of 'making anything better'. It is fundamentally a 'procedure of changing inputs' to accomplish 'either the least or the most extreme yield' for any recognized issue. The idea of optimization is fundamentally what we do in our day-to-day lives: a want to improve the situation or be best in some fields. In engineering, the basic concepts of optimization may be defined as follows: any process or methodology towards any design, system or even decision that may be fully/partially perfect, functional or effective as possible; specifically, in line of mathematical modelling or procedures involved in it.

In any implicational field, we wish to deliver the most ideal outcome with accessible assets. In a profoundly focused business world, it is not any more adequate to outline a framework whose execution of required errand is simply tasteful. The basic idea is to configure the best framework in a best optimal manner. Accordingly, by 'designing' new items in fields such as aviation, car, electrical, biomedical and horticultural, we endeavour to utilize design tools that give wanted outcomes in a convenient and prudent way.

Related to the field of engineering applications, optimization is considered as an extremely broad robotized design procedure. Using this strategy, it is critical to recognize analysis and design the same. Analysis is a procedure for deciding the response of determined framework to a certain mix of information parameters. Design, on the other hand, implies a procedure for characterizing any framework.

We move on to the world of modelling in its most basic terms. Optimization is a choice discipline of mathematical modelling that generally concerns to find the extreme (maxima and minima) of numbers, functions or systems.

Great ancient philosophers and mathematicians created its foundations by defining the choice of optimum (as an extreme, maximum or minimum) over several fundamental domains such as numbers, geometrical shapes, optics, physics and astronomy.

In a research view of observation, optimization is defined by different researchers in different ways at different times. A conventional view of optimization may be presented as follows:

> Individuals' want for flawlessness discovers articulation in principle of Optimization. It examines how to depict and accomplish what is Best, once one knows how to quantify and change what is great or terrible… hypothesis of Optimization incorporates quantitative investigation of optima and methods for discovering them.

In nature, the method of optimization moves from the concept of 'free of constraints' (unconstrained) to 'constraints' (constrained) one. If the design of any system is at a stage where no constraints are active, then the process of determining search direction and travel distance for minimizing the objective function that involves an unconstrained minimization algorithm. Of course, in such a case, one has to constantly watch for constraint violations during the move towards design space. All unconstrained optimization methods are of 'iterative nature', which begin from an 'initial trial solution' and moves to an optimum point in a sequential process. All these methods require an initial point X_1 to start and differ from one another: only methods generating point X_{i+1} (from X_i) and testing point X_{i+1} for optimality. Any constrained optimization problem can be cast as an unconstrained minimization problem even if constraints are active.

1.2 Single-Objective Mathematical Programming

The Problem of Optimization (either maximizing or minimizing) of an algebraic or transcendental function of one or more variables subject to some specific constraints is called mathematical programming problem or constraint optimization problem or precisely a single-objective mathematical programming problem (SOMPP), which may be stated mathematically as follows:

Determine the values of variable $x = (x_1, x_2, \ldots, x_n)$ that optimize function (called objective function or criterion function)

$$\text{Optimize } f(x) \tag{1.1}$$

Subject to constraints:

$$g_i(x) \leq 0 \forall i = 1,2,\ldots,m$$
$$h_j(x) = 0 \forall j = 1,2,\ldots,l \quad (1.2)$$
$$x \geq 0$$

where $f(x), g_1(x),\ldots,g_m(x), h_1(x),\ldots,h_l(x)$ are functions designated on an n-dimensional set and x is defined as a vector of n components $x_1, x_2,\ldots,x_n$. For both, objective function and constraints are linear, and SOMPP evolves into single-objective linear programming problem (SOLPP).

In illustrating SOLPP, let us consider a minimization problem. This implies no loss of generality since, if the objective is to maximize $F(x)$, this can be converted to an equivalent minimization problem by taking $f(x) = -F(x)$ Similarly, $g_i(x) \geq 0$ type constraints can be easily converted into $G_i(x) \leq 0$ type $\left(G_i(x) = -g_i(x)\right)$.

An ideal SOLPP is therefore represented as follows:

$$\text{Minimize } g_0(x) \quad (1.3)$$

Subject to constraints:

$$g_k(x) \leq a_i;\ \forall k = 1,2,\ldots,m$$
$$h_j(x) = b_j;\ \forall j = 1,2,\ldots,l \quad (1.4)$$
$$x_i \geq 0;\ \forall i = 1,2,\ldots,n$$

where $x = (x_1, x_2,\ldots,x_n)^T$ is a vector of decision variables.

Any vector x fulfilling every constraint of equation (1.4) is said to be a feasible solution to the problem. Accumulation of every single such arrangement shapes a feasible region. The aim of SOLPP is therefore to find a feasible solution $\bar{x}$ such that $f(x) \geq f(\bar{x})$ for each feasible point x. Here, $\bar{x}$ is termed an 'optimal solution' or 'solution' to the problem.

For the solution of SOLPP, a lot of mathematical techniques based on linearization, use of gradient, etc. are available in literature. Here, we illustrate some of the methods, preferably in deterministic as well as stochastic environments, which have been utilized to solve different engineering optimization problems. Optimization refers to either minimization or maximization and is interchangeable in general; maximization of any function f is equal to minimization of the opposite of the same function ($-f$).

The concept of optimization is very much symbiotic to computers and is an ideal instrument for application as long as a thought or variable affecting the thought could be an input in electronic or quantitative configuration. The basic idea is to identify the problem in hand, feed it in terms of optimization problem and go for a solution. For a solution, feed the computer with real-life data and come up with outcomes of solution to problem.

1.3 Generalized Reduced Gradient Method

This method is used for solving non-linear programming problems for taking care of correspondence and disparity limitations. Consider a non-straight programming issue:

$$\text{Maximize } f(x) \tag{1.5}$$

Subject to constraints:

$$\begin{aligned} &\phi_j(x) \le 0; \forall\left(j = 1,2,\ldots,m\right) \\ &\psi_k(x) = 0; \forall\left(k = 1,2,\ldots,l\right) \end{aligned} \tag{1.6}$$

Here, all n-components of decision vector x are non-negative finite numbers.

Through adding non-negative slack variables $s_j(\ge 0); \forall\left(j = 1,2,\ldots,m\right)$ to all of previous inequality constraints, the aforestated problem (1.5) could be obtained as

$$\text{Maximize } f(x) \tag{1.7}$$

Subject to constraints:

$$\begin{aligned} &\phi_j(x) + s_j = 0; \forall\left(j = 1,2,\ldots,m\right) \\ &\psi_k(x) = 0; \forall\left(k = 1,2,\ldots,l\right) \end{aligned} \tag{1.8}$$

Lower and upper bounds on slack variables $s_j\left(\forall j = 1,2,\ldots,m\right)$ are taken between zero to a large number considerably.

The mentioned problem could be rewritten as

$$\text{Maximize } f(x) \tag{1.9}$$

Subject to constraints:

$$\xi_j(x) = 0; \forall\left(j = 1,2,\ldots,m+l\right) \tag{1.10}$$

where the $(n+m)$ components of decision vector are non-negative finite numbers.

Strategy for generalised reduced gradient method (GRG) depends on the fundamental idea of disposal of factors by utilizing uniformity requirements. Hypothetically, $(m+1)$ subordinate factors can be communicated in terms of $(n-1)$ independent factors. Accordingly, one can separate choice

$(n+m)$ factors arbitrarily into two sets as $x=(y,z)^T$, where y is $(n-1)$ outline or free factors and z is $(m+l)$ state or independent factors, and

$$y=(y_1,y_2,\dots,y_{n-l})^T$$

$$z=(z_1,z_2,\dots,z_{m+l})^T$$

Outline factors, here, are totally autonomous. State factors are subject to outline factors used to fulfil imperatives: $\xi_j(x)=0;(\forall j=1,2,\dots,m+l)$

Consider first variations of objective and constraint functions as follows:

$$df(x)=\sum_{i=1}^{n-l}\frac{\partial f}{\partial y_i}\partial y_i+\sum_{i=1}^{m+l}\frac{\partial f}{\partial z_i}\partial z_i=\nabla_y^T f dy+\Delta_z^T f dz$$

$$d\xi_j(x)=\sum_{i=1}^{n-l}\frac{\partial \xi_j}{\partial y_i}\partial y_i+\sum_{i=1}^{m+l}\frac{\partial \xi_j}{\partial z_i}\partial z_i \tag{1.11}$$

or

$$d\xi=Cdy+Ddz \tag{1.12}$$

where

$$\nabla_y^T f=\left(\frac{\partial f}{\partial y_1},\frac{\partial f}{\partial y_2},\dots,\frac{\partial f}{\partial y_{n-l}}\right)$$

$$\nabla_z^T f=\left(\frac{\partial f}{\partial z_1},\frac{\partial f}{\partial z_2},\dots,\frac{\partial f}{\partial z_{m+l}}\right)$$

$$C=\begin{bmatrix}\frac{\partial \xi_1}{\partial y_1} & \frac{\partial \xi_1}{\partial y_{n-l}}\\ \frac{\partial \xi_2}{\partial y_1} & \frac{\partial \xi_2}{\partial y_{n-l}}\\ & \\ \frac{\partial \xi_{m+l}}{\partial y_1} & \frac{\partial \xi_{m+l}}{\partial y_{n-l}}\end{bmatrix},D=\begin{bmatrix}\frac{\partial \xi_1}{\partial z_1} & \frac{\partial \xi_1}{\partial z_{m+l}}\\ \frac{\partial \xi_2}{\partial z_1} & \frac{\partial \xi_2}{\partial z_{m+l}}\\ & \\ \frac{\partial \xi_{m+l}}{\partial z_1} & \frac{\partial \xi_{m+l}}{\partial z_{m+l}}\end{bmatrix}$$

$$dy=(dy_1,dy_2,\dots,dy_{n-l})^T$$

$$dz=(dz_1,dz_2,\dots,dz_{m+l})^T$$

Accepting that limitations are to be fulfilled at vector x for which $\xi(x)=0$, any adjustment in vector dx must relate to $d\xi = 0$ by keeping the possibility at $(x+dx)$. Along these lines, the previous equation can be solved as

$$\begin{aligned} &Cdy + Ddz = 0 \\ &\Rightarrow dz = -D^{-1}Cdy \end{aligned} \tag{1.13}$$

Change in objective function is due to progress in x; controlled by condition (1.4), which can also be reconstrained utilizing condition (1.13) as follows:

$$\begin{aligned} &df(x) = \left(\nabla_y^T f - \nabla_z^T fD^{-1}C\right)dy \\ &\Rightarrow df(x)/dy = G_R \\ &\Rightarrow G_R = \nabla_y^T f - \nabla_z^T fD^{-1}C \end{aligned} \tag{1.14}$$

is called a generalized reduced gradient. Geometrically, lessened gradient can be depicted as a projection of a unique n-dimensional gradient into $(n-m)$-dimensional feasible region portrayed by design factors.

The fundamental condition for the presence of any unconstrained capacity is that 'parts of gradient ought to vanish'. Additionally, obliged work expects its base esteem when proper segments of decreased gradient are zero. Actually, decreased gradient G_R can be utilized to create the course of search direct S toward lessen estimation of obliged objective function. Comparable is the case with gradient ∇f that can be utilized to produce a search course S for an unconstrained capacity. An appropriate advance length λ is to be limited estimation of $f(x)$ alongside search course. For a particular estimation of λ, subordinate variable vector z is refreshed using condition (1.13). Taking note that condition (1.12) depends on using a straight guess to a unique non-direct issue, we find that imperatives may not be precisely equivalent to zero at λ, i.e., $d\xi \neq 0$. Subsequently, when y is held settled, with a specific end goal to have

$$\xi_j(x) + d\xi_j(x) = 0; \left(\forall j = 1,2,\ldots,m+l\right) \tag{1.15}$$

We must have

$$\xi(x) + d\xi(x) = 0 \tag{1.16}$$

Using equation (1.12) for $d\xi$ in equation (1.16), we obtain

$$dz = D^{-1}\left(-\xi(x) - Cdy\right) \tag{1.17}$$

The value of dz is used to update the value of z as follows:

$$z_{\text{update}} = z_{\text{current}} + dz \tag{1.18}$$

Constraints assessed at refreshed vector x and system of discovering dz using condition (1.18) are rehashed until they are adequately little.

1.4 Derivative-Based Optimization

Here, we discuss a fundamental class of gradient-based optimization techniques, capable of determining search directions according to derivative information of any objective function. First, we start with preliminary concepts that sustain descent algorithms used for solving minimization problems, as well as their relevant procedures, starting with steepest descent method and Newton's method as the basis for many gradient-based algorithms. A class of gradient-based methods can be applied to optimize non-linear neuro-fuzzy models, thereby allowing such models to play a prominent role in the framework of soft computing (SC).

1.4.1 Descent Methods

In this section, our main focus is on minimizing a real-valued objective function E defined on an n-dimensional input space $\theta = [\theta_1, \theta_2, \ldots, \theta_n]^T$. Finding a (possibly local) minimum point $\theta = \theta^*$ that minimizes $E(\theta)$ is of primary concern. In general, a given objective function E may have a non-linear form with respect to an adjustable parameter θ. Due to the complexity of E, we often resort to an iterative algorithm to explore the input space efficiently. In iterative descent methods, the succeeding next point θ_{next} is determined by a step-down approach from the current point θ_{now} in a direction vector d:

$\theta_{\text{next}} = \theta_{\text{now}} + \eta d$; η is some positive step size regulating to what extent to proceed in that direction. This, again, can be alternately represented as follows:

$$\theta_{k+1} = \theta_k + \eta_k d_k \forall k = 1,2,3,\ldots$$

where k denotes the current iteration number, and θ_{now} and θ_{next} represent two consecutive elements in a generated sequence of solution candidates $\{\theta_k\}$. This θ_k is intended to converge to a (local) minimum θ^*.

Iterative descent methods compute kth step $\eta_k d_k$ through the following procedure:

Step 1: Determining direction d.

Step 2: Calculating step size η.

Step 3: The next point θ_{next} should satisfy the following inequality:

$$E(\theta_{\text{next}}) = E(\theta_{\text{now}} + \eta d) \prec E(\theta_{\text{now}})$$

1.4.2 Gradient-Based Method

Descent methods with straight downhill direction d, when determined on the basis of gradient (g) of an objective function E, are called gradient-based methods.

Gradient of differentiable functions $E: R^n \to R$, where θ is the vector of first derivatives of E, generally labelled as g.

$$g(\theta) = (= \nabla E(\theta)) \overset{\text{def}}{=} \left[\frac{\partial E(\theta)}{\partial \theta_1}, \frac{\partial E(\theta)}{\partial \theta_2}, \ldots, \frac{\partial E(\theta)}{\partial \theta_n} \right]^T$$

A class of gradient-based descent methods has the following fundamental form, in which feasible descent directions can be determined by deflecting gradients through multiplication by G and we have $\theta_{\text{next}} = \theta_{\text{now}} - \eta G_g$ with some positive step size η and some positive definite matrix G. Generally, we wish to find a value of θ_{next} which satisfies the following conditions:

$$g(\theta_{\text{next}}) = \left| \frac{\partial E(\theta)}{\partial \theta} \right|_{\theta = \theta_{\text{next}}} = 0$$

This is extremely hard to comprehend the previous condition logically. To limit the target work, plunge strategies are rehashed until the point that any of the halting criteria is fulfilled:

1. Value of target work is sensibly little.
2. Length of gradient vector g is littler than a pre-defined value.
3. Specified registering time is surpassed.

1.4.3 Steepest Descent Method

Steepest Descent Method (popularly known as gradient method) is an old technique used for minimizing any given function defined on a multi-dimensional input space. This method forms the basis for many direct methods used in optimizing both constrained and unconstrained problems. Despite its sluggish convergence capability, it is considered maximum regularly utilizing non-linear optimization technique because of its ease in applying several implicational areas. The well-known formula for evaluating the steepest descent formula is $\theta_{\text{next}} = \theta_{\text{new}} - \eta g$ with some positive step size η and gradient direction g.

Application of the steepest descent method may be diversified in many areas for practical optimization application. It can be employed as a touchstone for discovering intrinsic difficulty of a given task and for establishing a certain reference for performance comparison. Particularly, for a large and complex problem, it must be worthwhile to use it due to its simplicity.

1.4.4 Newton's Method (Classical Newton's Method)

Descent direction d can be determined through second derivatives of objective function E, if accessible. For a general continuous objective function, contours may be nearly elliptical in immediate vicinity of minimum; objective function E is expected to be approximated by a quadratic form:

$$E(\theta) \approx E(\theta_{\text{new}}) + g^T(\theta - \theta_{\text{now}}) + \frac{1}{2}(\theta - \theta_{\text{now}})^T H(\theta - \theta_{\text{now}})$$

if the starting position θ_{next} is sufficiently close to a local minimum. Here, H is a Hessian matrix, consisting of second partial derivatives of $E(\theta)$. The previous equation is a quadratic function of θ; we can simply find its minimum point $\hat{\theta}$ by differentiating the same and setting it to zero. Finally, this leads to a set of linear equations like $0 = g + H(\hat{\theta} - \theta_{\text{now}})$. If an inverse of H exists, we have a unique solution. When the smallest point $\hat{\theta}$ of the estimated quadratic function is selected as the following point θ_{now}, we consume the so-called Newton's technique expressed in the following form:

$$\hat{\theta} = \theta_{\text{new}} - H^{-1}g$$

$-H^{-1}g$ is called Newton step, and its direction is called Newton direction. General gradient-based formula $\hat{\theta} = \theta_{\text{new}} - H^{-1}g$ is reduced to Newton's method when $G = H^{-1}$ & $\eta = 1$. If H is positive definite and $E(\theta)$ is quadratic, at that point, Newton's method directly gets to a local minimum in a single Newton phase. If $E(\theta)$ is not quadratic, then minimum might not be touched in a solitary stride, and Newton's method should be repeatedly employed. For practical purpose, the usual practice is to modify Newton's method as evaluating Hessian may not be worthwhile due to its heavy computational requirements. Many promising algorithms can be regarded as a sort of intermediate between steepest descent and Newton methods (e.g., conjugate gradient methods and Levenberg–Marquardt methods, which are widely used and considered as good general-purpose algorithms for solving non-linear least-squares problems).

1.5 Soft Computing

Soft computing (SC) is basically a technique of computational philosophy used to handle problems related to interdisciplinary areas such as computer science and complex engineering disciplines. It endeavours to think about, show and break down extremely complex wonders: those for which customary techniques have not yielded minimal effort, logical and finish arrangements. Prior computational methodologies by and large could show and unequivocally dissect just straightforward frameworks. Complex frameworks in science, medication, humanities, administration sciences, computerized reasoning and so on, are usually unmanageable by ordinary numerical and investigative techniques.

SC is considered as a heap of deliberate techniques that were confined to demonstrate and empower answers for genuine issues – either not displayed or excessively troublesome, making it impossible to show, as communicated scientifically. In these angles, it is a consortium of techniques that work synergistically and give, in some frame, adaptable data-preparing capacity for taking care of genuine uncertain circumstances. Regarding execution, the point of this technique is to misuse resilience for imprecision, vulnerability, rough thinking and incomplete truth to accomplish tractability, power and minimal effort arrangements. The controlling rule of SC is used to devise strategies for calculation that prompt a satisfactory 'best ideal arrangement' requiring little to no effort or by looking for an estimated answer for a loosely or exactly figured issue of concern.

SC, unlike traditional/customary (hard) computing, works on systems containing fuzziness, vulnerability, incomplete truth and estimate. As a result, a good example for SC is to map with human personality. Standard directing to SC remains to exploit resilience intended for imprecision and vulnerability and to accomplish tractability, vigour and little arrangement cost. The essential idea of simple SC is its present incarnation to have connections to numerous prior impacts, amongst them Zadeh's commitment to fuzzy sets in 1965, and write about the probability hypothesis and soft information examination in 1979 and 1981. Before 1994 when Zadeh first defined 'soft computing', he took care of ideas used to be alluded to in a separate way, whereby everyone talked about exclusively with a sign of utilization of fuzzy systems. Despite the fact that the idea of establishing up the area of SC dates back to 1990, Prof. Zadeh defined SC as follows:

> Basically, soft computing is not a homogeneous body of concepts & techniques. Rather, it is a partnership of distinct methods that in one way or another conform to its guiding principle. At this juncture, dominant aim of soft computing is to exploit tolerance for imprecision & uncertainty to achieve tractability, robustness & low solutions cost. Principal constituents of soft computing are fuzzy logic, neuro-computing, & probabilistic reasoning, with latter subsuming genetic algorithms, belief networks, chaotic systems, & parts of learning theory. In partnership of fuzzy logic, neuro-computing, & probabilistic reasoning, fuzzy logic is mainly concerned with imprecision & approximate reasoning; neuro-computing with learning & curve-fitting; & probabilistic reasoning with uncertainty & belief propagation'.

Guiding principle, as per Prof. Zadeh regarding soft computing is:

> 'Exploit tolerance for imprecision, uncertainty, partial truth, & approximation to achieve tractability, robustness & low solution cost'.

Inclusion of neural and genetic computing in domain of SC came at a later point.

1.5.1 SC – How Is It Different from Hard Computing?

Human thinking is transcendently approximated, subjective and 'soft'. People can adequately deal with fragmented, uncertain and fuzzy data in settling on shrewd choices. Fuzzy rationale, likelihood hypothesis, neural systems and genetic calculations are agreeably utilized as a part of it for learning portrayal and for imitating thinking and basic leadership process. SC is an interdisciplinary approach in intersection territory of shrewd and information-based frameworks. It has successfully supplemented ordinary artificial intelligence in the territory of machine knowledge (computational insight). SC concerns utilization of speculations of fuzzy rationale, neural systems and evolutionary computing to take care of genuine issues that cannot be acceptably unravelled utilizing customary fresh computing procedures and represents a set of computational intelligence-based methodologies that are modelled to deal effectively with systems characterized by complex structures and possibly subjected to incomplete knowledge and ill-defined dynamics.

Hard computing, likewise named as ordinary computing, requires a decisively expressed explanatory model and frequently requires a ton of calculation time. Hard computing depends on binary rationale, fresh frameworks, numerical investigation and fresh software though the base for SC deals with fuzzy rationale, neural nets and probabilistic thinking. Hard computing takes the attributes of accuracy and categoricity, while SC works better with qualities such as estimate and dispositionality. In tough calculation, imprecision and vulnerability are bothersome properties, wherein in SC, resistance intended for imprecision and vulnerability is misused to accomplish controllability, high Machine Intelligence Quotient and economy of correspondence. Hard computing expects projects to be composed, whereas SC could develop its own projects. Hard computing utilizes two-esteemed rationale, whereas SC utilizes multi-esteemed or fuzzy rationale. Hard computing is, as a rule, deterministic, whereas SC is stochastic. Hard computing needs correct information, whereas SC manages questionable and boisterous information. Hard computing is entirely consecutive in process, whereas SC permits parallel calculations. Finally, while hard computing yields exact responses, SC knows how to produce surmised responses as it were. In situations where the exact answer is greatly intense, soft calculation uses a better route for determining the 'best ideal' arrangement. The general distinction is very much featured by Prof. Lotfi Zadeh:

> Soft computing differs from conventional computing in that, unlike hard computing, it is tolerant of imprecision, uncertainty, partial truth, & approximation. In effect, role model for soft computing is human mind interface.

1.6 Relevance of SC towards Optimal Solution

Hard computing requires decisively expressed diagnostic model and requires parcel of calculation time. While analytical models are fit/legitimate for perfect cases, certifiable issues exist in non-perfect condition. At this point, principal components of SC are termed as Machine Learning, Fuzzy Logic, Evolutionary Computation, Neural Computing and Probabilistic Reasoning, with the latter including chaos theory belief networks and parts of learning theory. Ideas of SC are not a concoction; reasonably, it is an association in which all of the accomplices underwrite a particular technique aimed at tending to issues in its space. Through this point of view, vital constituent philosophies in SC are coordinated more correlative as opposed to aggressive.

As SC is considered as a developing gathering of interdisciplinary techniques, it has been worthwhile in numerous genuine applications. As opposed to investigative strategies, SC techniques impersonate cognizance and comprehension in a few critical regards: they can gain as a matter of fact; they can universalize into areas where coordinated experience is not found when all is said and done; and through parallel computer architectures that mimic natural procedures, they can perform mapping from contributions to yield speedier than inalienably serial explanatory portrayals. Exchange off, be that as it may, is a decline in exactness. In the event that an inclination towards imprecision could be endured, at that point it ought to be conceivable to stretch out the extent of utilizations even to those issues where investigative and scientific portrayals are promptly accessible. Inspiration for such an expansion is a normal decline in computational load and resulting increment of calculation speeds that allow heartier framework.

There have been different ensuing endeavours to additionally sharpen part of SC towards optimization; maybe the most reasonable is as follows:

> Each computing procedure that intentionally incorporates imprecision into count on at least one levels and permits this imprecision also to modification (diminish) granularity of issue, or to 'soften' objective of optimization at some phase, is characterized as to having a place with field of soft computing.

Perspective which we will ponder here is another method for characterizing SC; therefore, it is thought towards the direct opposite of what we may call hard computing. SC may well along these lines be viewed as a grouping of procedures and strategies to apply to genuine handy circumstances in the same route as human would tackle them. In this logic, SC is a group of issue determination techniques controlled by inexact thinking and useful and optimization estimation strategies, with search strategies. SC is, in this manner, considered as a hypothetical reason for the area of smart frameworks, and it is apparent that contrast between the zone of computerized reasoning and that of clever frameworks is that it initially depends on hard computing and then on SC.

Commencing this perspective on a moment level, SC could be ventured in different segments that add to a definition by augmentation, for example, the one initially given. From the beginning, segments thought to be utmost vital in this second level are probabilistic thinking, fuzzy rationale and fuzzy sets, neural systems and genetic calculations, which on account of their interdisciplinary uses and outcomes promptly emerged over different approaches, for example, the aforementioned chaos theory, evidence theory etc. Prevalence of Genetic Algorithm, together through their validated proficiency in a wide assortment of territories and uses, their endeavour to mimic normal animals which are obviously soft and particularly augmentations and diverse variants, change this fourth second-level fixing into surely understood evolutionary calculations.

Real-world problems have to deal with systems that are non-linear, time-varying in nature with uncertainty and high complexity. Computing of such systems is the study of algorithmic methodologies that describe and transform information: its theory, design, efficiency, implementation, analysis and application. Conventional/hard computing requires the exact mathematical model and lot of computation time. For such problems, methods which are computationally intelligent possess human-like expertise, can adapt to changing environmentand can be used effectively and efficiently. SC is an evolving collection of artificial intelligence methodologies aiming to exploit tolerance for imprecision and uncertainty that is inherent in human thinking and in real-life problems, to deliver robust, efficient and optimal solutions and to further explore and capture available design knowledge. It utilizes computation, reasoning and inference to reduce the computational cost by exploiting tolerance for imprecision, uncertainty, partial truth and approximation. Numerous SC-based methods and applications have been reported in literature in a range of scientific domains. Advances in application of SC techniques in various demanding domains have promoted its use in industrial applications. SC with its roots in fuzzy logic, artificial neural network and evolutionary computation has become one of the most important research fields applied to numerous engineering areas such as aircraft, communication networks, computer science, power systems and control applications.

SC technology is of great importance for data compression, especially in high definition television, audio recording, speech recognition, image understanding and related fields. Techniques of SC consist of rich knowledge representation, knowledge acquisition and knowledge processing for solving various applications. These techniques can be deployed as individual tools or be integrated in unified and hybrid architectures. Fusion of these techniques causes a paradigm shift in engineering and science fields, which could not be solved with conventional computational tools. With wide applicability, SC has gained special importance in application fields for wireless communication. Wireless communication systems are associated with much uncertainty and imprecision due to a number of stochastic processes, such as time-varying characteristics of wireless channel caused by mobility of transmitters, receivers, objects in environment and mobility of users. This reality has fuelled numerous applications of SC techniques in mobile and wireless communications, and the role of SC in domain of wireless systems can be classified into three broad categories: optimization, prediction and uncertainty management.

Fruitful utilizations of SC combined with AI propose that it will have progressively more prominent effect in the coming years. SC is now assuming a critical part in both science and designing. In many ways, SC speaks to a huge change in outlook (achievement) in the point of computing, a move that resembles human personality, not at all like condition of – craftsmanship PCs, has an amazing capacity to store and process data, which is inescapably loose, unverifiable and ailing in categoricity. SC can be stretched out to incorporate computing not just from human speculation perspectives (mind and cerebrum) but also from bio-informatics viewpoints. As such, subjective and responsive appropriated man-made brainpower will be created and connected to huge-scale and complex mechanical frameworks. In fuzzy frameworks, calculation with words will be explored progressively, and additionally, evolutionary calculation will develop. It is normal that they will be connected to the development of further developed wise mechanical frameworks.

SC is a noteworthy territory of scholarly research. In any case, the idea is as yet advancing, and new philosophies, for example, confusion computing and invulnerable systems, are nowadays considered to have a place in this territory. While this methodological advancement is occurring, a number of effective SC-based items are expanding simultaneously. In larger part of such items, SC is covered up inside frameworks or sub-systems, and the end client does not really realize that SC techniques are utilized as a part of control, conclusion, design acknowledgement, flag preparing and so forth. This is the situation when SC is for the most part utilized for enhancing the execution of customary hard computing calculations or notwithstanding supplanting them. Be that as it may, SC is exceptionally powerful when it is connected to certifiable issues that are not ready to be tackled by customary hard computing. Another class of items utilizes SC for actualizing novel, keen and easy-to-understand highlights. SC empowers mechanical frameworks to be inventive because of imperative qualities of SC: tractability, high machine knowledge remainder and ease.

It is, hence, evident that in contrary to the exact meaning for SC, it is rather characterized by augmentation, by methods for various ideas and strategies which endeavour to face challenges which emerge in genuine issues happening in a world which is loose, indeterminate and hard to arrange. As an improvement of SC advances in a few orders including physical science, biological science and material science, PC researchers must know about their parts and prepare themselves for more prominent progression of SC in not so distant future.

2

Various Soft Computing Techniques and Their Description

2.1 History of Heuristics/Search Algorithms

Heuristics, prevalently known as estimated arrangement strategies, have been utilised since the beginning of operations research to handle troublesome combinatorial issues. With the improvement of many-sided quality hypothesis in the mid-1970s, it turned out to be evident that, since the majority of these issues were surely non-deterministic polynomial-time (NP)-hard, there was little desire for consistently finding productive correct arrangement methods for them. This acknowledgement accentuated part of heuristics for taking care of combinatorial issues that were experienced, all things considered and applications that should have been handled, regardless of whether they are NP-hard. Though various methodologies are anticipated and different things are tried, the major mainstream one depended on local search (LS) changes procedures. This can be generally compressed as an iterative pursuit system that, beginning from an underlying plausible arrangement, continuously enhances it by applying a progression of nearby adjustments (or moves). At every emphasis, look moves to an enhancing plausible arrangement that contrasts just somewhat from the current one. Seek ends when it experiences a neighbourhood ideal concerning changes that it considers a vital restriction of strategy: unless one is to a great degree fortunate, this nearby ideal is frequently a genuinely average arrangement. In LS, the nature of arrangement acquired and computing times are generally an exceptionally endless supply of set of developments considered at every cycle of heuristic.

In 1983, the universe of combinatorial optimization was broken by advent of ideas of another heuristic approach termed simulated annealing (SA) that can be appeared to meet to an ideal arrangement of a combinatorial issue, yet in boundless computing time. In view of similarity with factual mechanics, SA can be deciphered as a type of controlled arbitrary stroll in space of practical arrangements. Rise of SA showed that one could search for different approaches to handle combinatorial optimisation issues and have prodded enthusiasm of the research group. In the following years, numerous other methodologies, generally in view of analogies with normal wonders, were proposed (Tabu Search, Ant Systems, Threshold Methods and so on.) and, together with more established ones, for example, Genetic Algorithms (GAs), they picked up an expanding prominence. Presently on the whole, known under the name of Meta-Heuristics, these methods have turned out to be over the most recent 15 years driving the edge of heuristic methodologies for taking care of combinatorial optimisation issues.

2.2 Soft Computing Approaches – Derivative-Free Optimisation

We discuss about the most prominent subordinate unrestricted optimisation methods: GA, SA, Random search technique and Downhill simplex search which are widely utilized to comprehend both ceaseless and discrete optimisation issues. Normal qualities of these populace-based stochastic search methods are as follows:

Derivative freeness – These methods do not require useful subsidiary data to search for an arrangement of parameters that limit (or amplify) a given target work.

Intuitive guideline – Guidelines followed by these search methods are generally in the light of basic instinctive ideas: once in a while propelled by nature's move, for example, evolution and thermodynamics.

Slowness – Without using derivatives, these methods are bound to be generally slower than derivative-based optimization methods for continuous optimization.

Flexibility – Derivative freeness likewise soothes the necessity for differentiable target capacities, so we can use as intricate a target work as a particular application may require, without giving up excessively in additional coding and calculation time.

Randomness – All these methods (with likely special case of standard downhill simplex search) are stochastic, which implies that they all utilize arbitrary number generators in deciding consequent search bearings. These components of haphazardness for the most part offer ascent to excessively hopeful view that these methods are 'worldwide analysers' that will locate a worldwide ideal sufficiently given calculation time.

Analytic Opacity – It is hard to do expository investigations of the named methods, to a limited extent as a result of their arbitrariness and issue –particularly nature.

Iterative nature – Unlike straight slightest squares estimator, these methods are iterative in nature. We require certain stopping criteria to choose when to end optimization process.

Let k indicate an iteration check and signify the best objective function obtained at tally k; regular halting criteria for a maximization issue include the following:

- *Computational time*: time required for calculation.
- *Optimisation objective*: f_k is not as much as a specific preset objective esteem.
- *Minimal change*: $\left(f_k - f_{k-1}\right)$ is not as much as a preset esteem.
- *Minimal relative change*: $\frac{\left(f_k - f_{k-1}\right)}{f_{k-1}}$ is not as much as a preset esteem.

As of late, both GAs and SA had been expanding the measures of consideration because of their adaptable optimization abilities in explaining both non-stop and discrete optimization issues. In addition, the two are propelled by supposed nature's astuteness: GAs are approximately in view of ideas of characteristic choice and evolution, while SA began in annealing forms found in thermodynamics. Random search and downhill simplex search is primarily used for continuous optimisation problems. Here, we will discuss all derivative-free advanced approaches that are more flexible in terms of incorporating intuitive guidelines and forming sophisticated objective functions in dealing with engineering and management science problem.

2.2.1 Genetic Algorithms

In the 1950s and 1960s, a few computer researchers freely contemplated evolutionary systems, having thought that evolution could be utilized as an optimization technique for designing issues. Thought in every one of these systems was to develop a populace of applicant answers for a given issue, utilizing administrators motivated by common genetic variety and normal determination. A GA was imagined by John Holland in 1960s and was created by Holland and his understudies and associates at University of Michigan amid this period. Conversely with evolution procedures and evolutionary writing computer programs, Holland's unique objective was not to plan algorithms to take care of particular issues, but instead to formally think about marvel of adjustment as it happens in nature and to create courses in which instruments of normal adjustment may be incorporated into computer systems. Holland's GA is a technique for moving from one populace of 'chromosome' to another populace by utilizing a sort of 'regular determination' together with genetics-enlivened administrators of hybrid, transformation and reversal. Focal thought behind it is power; adjust between productivity and viability important for survival in a wide range of conditions. It is a subordinate-free stochastic optimization technique construct freely with respect to ideas of common choice and evolutionary procedures. A GA is a type of evolution that occurs on a PC and is a computational model propelled from standards of common evolution and populace genetics. It is probabilistically guided optimisation methods and is utilized for both taking care of issues and demonstrating evolutionary systems. It is a search algorithm that joins 'survival of fittest' amongst string structures having organized yet randomized data trade to frame a search algorithm by means of some of imaginative pizazz of human search strategies. Persuaded by Darwin's hypothesis of evolution and idea of 'survival of fittest', GAs utilize forms practically equivalent to genetic recombination and change to advance evolution of a populace that best fulfils a pre-defined objective. As a universally useful optimization device, GAs are moving out of the scholarly world and verdict noteworthy uses in numerous different settings. Their fame could be credited to their opportunity from reliance on practical subsidiaries and to their joining of these attributes:

1. GAs are parallel-search strategies that may be actualized on parallel-handling machines for enormously accelerating their operations.
2. They are appropriate to both constant as well as discrete (combinatorial) optimization issues.
3. They are stochastic and more averse to get caught in nearby minima that are definitely available in any down-to-earth optimization use.
4. Their adaptability encourages both structure and parameter distinguishing proof in complex models, for example, neural systems and fuzzy derivation systems.

GAs encode each point in a parameter (or arrangement) space into a binary bit string termed chromosome, and every point is related with a fitness esteem, that, for minimization, is normally equivalent to target work assessed at a point. Rather than a solitary point, GAs typically keep an arrangement of focuses as a populace, which is then advanced more than once towards a superior general wellness esteem. In every age, GA develops another populace utilising genetic administrators, for example, hybrid and transformation; individuals with higher wellness esteems will probably survive and partake in mating (hybrid) operations. After various ages, populace contains individuals with better wellness esteems; this is undifferentiated from Darwinian models of evolution by arbitrary change and regular determination. GAs and their variants are once in a while alluded to as approaches of populace-grounded optimization that enhances execution by redesigning whole populaces as opposed to singular individuals. Real parts of GAs incorporate encoding plans, wellness assessments, parent determination, hybrid administrators and transformation administrators. The end is a prefix with a few criteria. Elitism is one such property that is connected to a genetic operation because of strength with great chromosomes.

2.2.1.1 Encoding Schemes

These change focuses in parameter space into bit string portrayals. Other encoding plans (for example, Gray coding) can likewise be utilized when essential; game plan may be prepared for encoding negative, coasting point or discrete-esteemed numbers. Encoding plans give a method for deciphering issue particular information straightforwardly into the structure of GA, and accordingly assume a key part in deciding the execution of GAs. Genetic administrators, for example, hybrid and transformation can and ought to be outlined alongside encoding plan utilized for a particular application.

2.2.1.2 Fitness Evaluation

Initial step in the wake of making an age is to figure fitness estimation of every part in populace. For an amplification issue, fitness estimation of f_i ith part is typically a target work assessed at this part/point. We ordinarily require fitness esteems that are certain, so some sort of monotonically scaling and/or interpretation might be essential if target work isn't entirely positive. Alternative method is to utilize rankings of individuals in a populace as their fitness esteems. Its benefit is that target work should not be precise as long as it can give revise positioning data.

2.2.1.3 Selection

After assessment, we need to make another populace from current age. Selection operation figures out which guardians participate in delivering child, and it is undifferentiated from survival of fittest in characteristic choice. There is a distinctive selection system: Rank-based fitness assignment, Stochastic Universal Sampling, Roulette wheel selection, Truncation selection and Tournament selection to name a few. Normally, individuals are chosen for mating with a choice likelihood relative to their fitness esteems. Utmost basic approach to execute this is to set selection likelihood equivalent to $f_i/\sum_{k=1}^{n} f_k$, where n is a populace measure. Impact of this selection strategy is to permit individuals with better than expected fitness esteems to recreate and supplant individuals with beneath normal fitness esteems.

Complete steps used to perform selection process are as follows:

a. Calculate fitness value $\text{EVAL}(X_n)$ for every chromosome:

$$X_n(n = 1,2,\ldots,\text{POPSIZE})$$

b. Calculate selection probability p_n for each chromosome X_n:

$$p_n = \frac{\text{EVAL}(X_n)}{\sum_{\xi=1}^{\text{POPSIZE}} \left(\text{EVAL}(X_\xi)\right)}$$

c. Calculate cumulative probability q_n for each chromosome:

$$X_n : q_n = \sum_{\xi=1}^{n} p_\xi$$

d. Produce a random real number r in (0,1).
e. If $r < q_1$, then the first chromosome is X_1, or select nth chromosome:

$$X_n(2 \leq n \leq \text{POPSIZE}) \text{ such that } q_{n-1} < r \leq q_n$$

f. Repeat steps (d) and (e) POPSIZE times and find POPSIZE copies of chromosome.

2.2.1.4 Crossover

To misuse capability of current quality pool, we utilize hybrid administrators to produce new chromosomes that we expect will hold great highlights from the past age. Hybrid is generally connected to choose sets of guardians with a likelihood equivalent to a given hybrid rate. Contingent upon portrayal of factors, diverse methods must be utilized:

a. *Real-valued crossover*– Under this type of crossover, the following types are used by researchers: arithmetic crossover, simple crossover and heuristic crossover.
b. *Binary-valued crossover*– Under this type of crossover, the following types are used by researchers: single-point/double-point/multi-point crossover and uniform crossover.

Once more, on the premise of various sorts, *single-point crossover* is utmost essential crossover administrator, where a crossover point on genetic code is chosen aimlessly and two parent chromosomes are exchanged now. In *double-point crossover,* two crossover points are chosen and part of chromosome string among these double points is then exchanged to create two offspring. We may expand and characterize n-point crossover correspondingly. By and large, single-point crossover is an exceptional instance of n-point crossover. Impact of crossover is like that of mating in normal evolutionary process, in which guardians pass sections of their own chromosomes on their offspring. Along these lines, a few offspring can beat their folks on the off chance that they get 'great' qualities or genetic characteristics from the two guardians.

Algorithm for crossover is characterized in following manner:

a. First, produce a random real number r in (0, 1).
b. Furthermore, choose two chromosomes X_k and X_l randomly amongst population for crossover if $r <$ PXOVER.
c. Two offspring X_k' and X_l' are produced by crossover technique.
d. Repeat steps (a)–(c) POPSIZE/two times.

2.2.1.5 Mutation

By mutation, individuals are randomly altered. Crossover abuses current quality possibilities, yet in the event that populace does not contain all encoded data expected to take care of a specific issue, no measure of quality blending can create an agreeable arrangement. Mutation attempts to bump population gently into a slightly better course. This means that mutation changes single or all genes of a randomly selected chromosome slightly. There are different types of methods for mutation: (1) Real-valued mutation: Boundary mutation, Uniform mutation and Non-uniform mutation, and (2) Binary mutation.

The most basic method for executing mutation is to flip a bit through a likelihood equivalent to a low given mutation rate. A mutation administrator may keep any single bit from uniting to an incentive all through the whole populace and keep populace from focalizing and stagnating at any neighbourhood optima. Mutation rate is typically preserved low, so great chromosomes obtained from crossover are not misplaced. In the event that mutation rate is high, execution of GAs will approach that of a primitive arbitrary search.

Steps for performing mutations in GAs are as follows:

a. Initially, we produce a random real number r in (0, 1).
b. Then, we handpick a chromosome X randomly from population if $r <$ PMUT.
c. Third, we select a particular gene x_i of selected chromosome x randomly.
d. Then, a new gene x_i' corresponding to x_i is created by mutation technique.
e. Repeat steps (a)–(d) POPSIZE times.

2.2.1.6 Termination

Termination condition is a condition that stops the algorithm when either of the following three conditions is satisfied:

a. The best individual does not improve over 250 generations.
b. Total improvement of the last 20 best solutions is less than 0.001 (a small amount).
c. The number of generations reaches a maximum of 5,000.

As GAs are an iterative process, a typical practice is to end afterward a pre-defined number of ages and then look at best chromosomes in populace. In the event that no palatable arrangement is discovered, at that point these algorithms might be restarted.

In regular evolutionary process, choice, crossover and mutation all happen in a single demonstration of producing posterity. Here, we recognize among them obviously to encourage execution of and experimentation with GAs.

2.2.1.7 Elitism

Again for a convoluted application, to enhance execution of GA, we may pick an approach of continually keeping a specific number of best individuals when each new populace is created, which is called elitism. It is an expansion to numerous choice methods that powers GA to hold few numbers of best people at every age. Such people can be lost on the off chance that they are not chose to recreate or on the off chance that they are decimated by crossover or mutation.

In the light of previously mentioned ideas, a basic GA is portrayed in the following way:

Stage 1: Initialize a populace with arbitrarily created people and assess fitness estimation of every person.

Stage 2:

a. Chose two individuals from populace with probabilities corresponding to their fitness esteems.
b. Put on crossover with a likelihood equivalent to crossover rate.
c. Apply mutation with a likelihood equivalent to mutation rate.
d. Repeat steps (a)–(d) till the point when sufficient individuals are created to from people to come.

Stage 3: Repeat steps (2) and (3) till the point that a ceasing paradigm is met.

2.2.1.7.1 *Varieties of GA – Modified GA*

Modification with respect to classical GA is that in mod GA, instead of performing selection step 'select $P(t)$ from $P(t-1)$', we rather select independently r chromosomes for reproduction and r chromosomes to die.

Algorithm for mod GA

```
Begin
t←0
initialize Population P (t)
   evaluate Population P (t)
   while (not terminate - condition) do
begin
t←t+1
select parents from Population P (t-1)
select deeds from Population P (t-1)
       form Population P (t):  reproduce parents
evaluate Population P (t)
    end
end
```

2.2.1.7.2 *Contractive Mapping GA*

Convergence of GAs is one of the major difficult hypothetical issues in evolutionary computation (EC) area. Choice of initial population may influence only convergence speed. Yet, it is possible that no new population is accepted for a long time and algorithm loops trying to find a new population $P(t)$.

Algorithm for Contractive Mapping GA

```
begin
t←0
initialize Population P (t)
evaluate Population P (t)
   while (not terminate - condition) do
begin contractive mapping f(P(t))→P(t+1)
t←t+1
select parents from Population P (t-1)
recombine Population P (t)
evaluate Population P (t)
        if eval(P(t))≥eval(P(t))
        then t←t-1
end
end
```

2.2.1.7.3 *GA – An Object-Based Approach*

Class of optimization problems that may be addressed by GAs may be formulated as follows:

Optimize an objective function

$$f(x) \tag{2.1}$$

$$f : R^k \rightarrow R$$

$$x = [x_1, x_2, \ldots, x_k] \in R^k$$

such that

$$x_i \in D_i = [a_i, b_i]; i = 1, 2, \ldots, k$$

Domain of problem (solution space) is defined by

$$D = D_1 \cdot D_2, \ldots, D_k \quad (2.2)$$

Constraints other than domain limits of x may be introduced as penalties in objective function. Now, classifications of optimization methods are given next. Various optimization techniques available may be broadly classified as Calculus-based techniques, Random-guided search techniques and Enumeration techniques. Two main random-guided search techniques are SA and evolutionary techniques. Later, it further contains GAs and Evolutionary Programming.

2.2.1.7.4 *GA – How It Works?*

Algorithm begins by creating an underlying populace of arbitrary hopeful arrangements. Every person in populace is then granted a score in view of its execution. People with best scores are guardians and are sliced and joined together to influence youngsters as found in the Figure 2.1 flowchart. To add some decent variety to the populace, some arbitrary mutations are connected. These youngsters are scored, with best entertainers liable to be guardians in people to come. Eventually, the process is ended, and the best scoring individual in populace is taken to be a 'best' one.

Stepwise procedure involved in GA can be stated as follows:

```
begin
Choose an initial random/biased population.
Evaluate Population.
repeat
Select Mate Pool (Reproduction)
Perform cross - over & Mutation
Evaluate Population
end   when stopping condition is reached.
end
```

Figure 2.1 shows the flowchart drawn with the preceding steps.

Figures 2.2 and 2.3, respectively, show the flow diagram and GA cycle in terms of Genotype and Phenotype traits of Population.

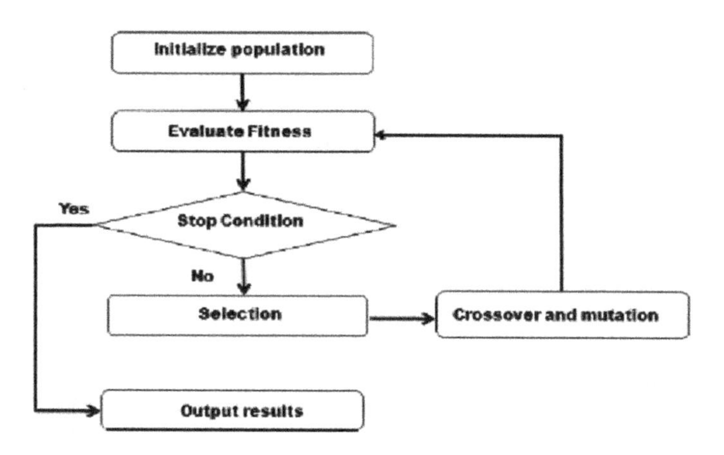

FIGURE 2.1
Flowchart of genetic algorithm.

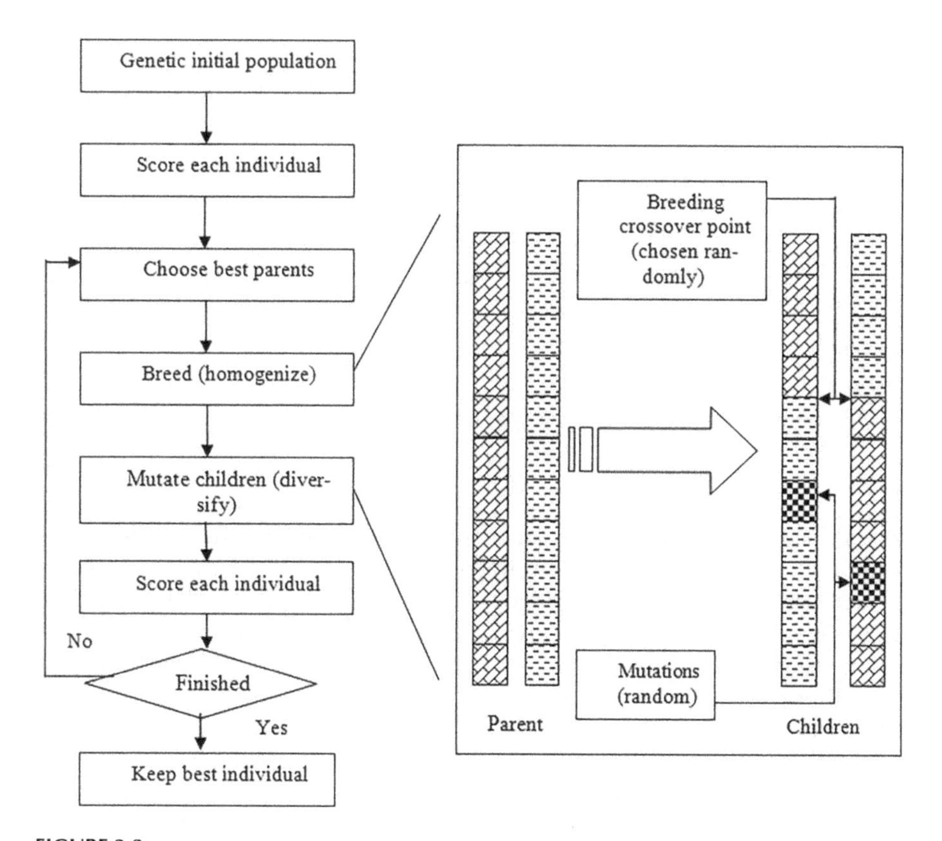

FIGURE 2.2
Flow diagram of genetic algorithm.

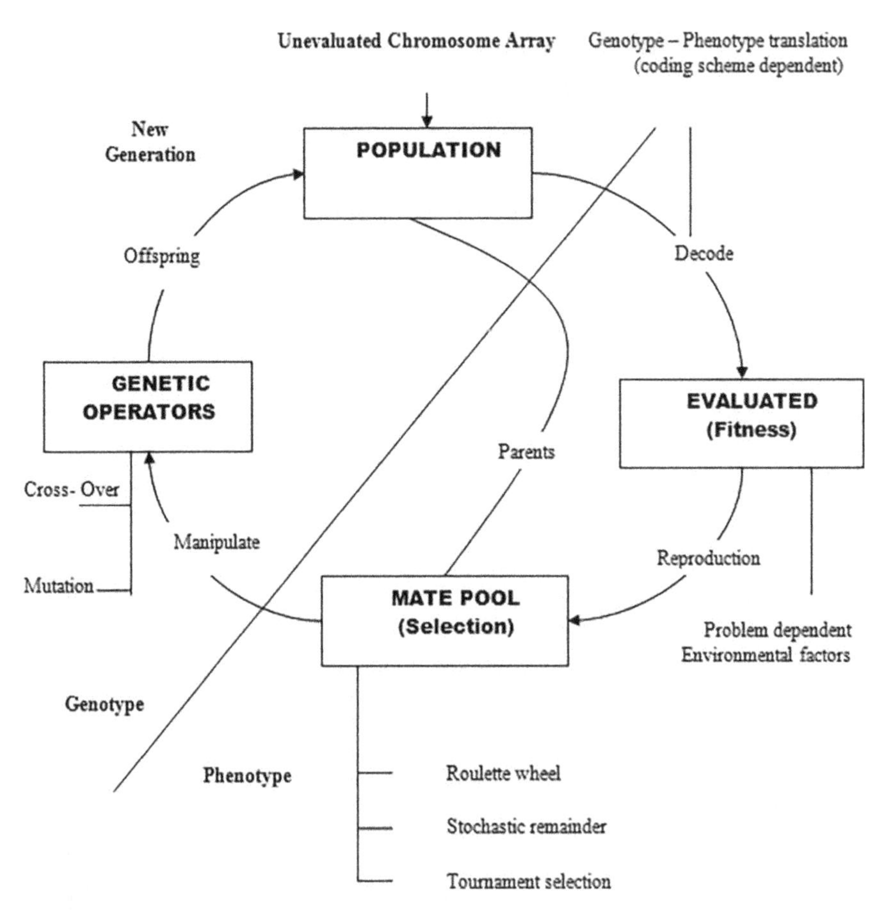

FIGURE 2.3
GA cycle in terms of genotype and phenotype traits of population.

2.2.1.7.5 Object Model of GA

As they stand, GA has only been looked upon from functional and computational points of view. Their implementation as algorithms has been concentrated upon since their inception. Here, an object model providing integrated views of structural, functional and computational aspects of GAs and emulating natural genetics on one-to-one basis at all levels is proposed. The following objects have been identified for design implementation of an object model:

1. *Gene* is a basic building block of natural genetics. Each gene can take one of the allelic values from a set, which is (0, 1) in case of binary-coded GAs. Hence, bit is the most suitable representation of a gene.

2. *Substring* is an array of genes. Each substring, of finite length l_i, represents a genotype of element x_i, which in turn is conceptually an attribute of an individual. An encoding mechanism has to be used to linearly map attribute values to gene strings. The length depends upon the extent of domain and precision requirements of the problem.
3. *Chromosome* is a cascade of substrings. This represents the genotype of an individual. The total length of a chromosome is $L = \sum_i l_i$.
4. *Individual* is an object that may be looked upon as a phenotypic manifestation of a chromosome. Each individual is, therefore, associated with vector x in problem domain and a fitness value $f(x)$, which decides the selection of an individual for manipulation.
5. *Population,* as mentioned previously, is a multi-set of individuals and represents a small subset of potentially optimum solutions in an entire search space. All genetic manipulation operators are defined on population. Genetic cycle ends when the majority of population converges to a seemingly global value.
6. *Mate-pool* is an object defined to encompass all operations such as shuffling and roulette wheel simulation, and to implement various selection algorithms to select a mate pool for current population.
7. *Generation* is an object defined to contain two population instances. One is current population and the other is a temporary population into which manipulated individuals of the current population will go.

   Other components of this object include statistical information and mate pool structure needed during each GA cycle. The object model so developed has one-to-one correspondence with natural genetics at structural as well as functional level. For implementation purposes, the following class constructs, too, have been identified to complete object model design.
8. *Statistics* are computed for each generation. These include parameters such as best fitness, average fitness, best individual, number of crossovers and mutation in last generation. This information is used for convergence analysis and tuning of GA for specific problem and also for determination of termination condition.
9. *Substring-Parameter* is an object that contains information regarding the domain of each substring (lower and upper bounds) and length of substring. This information is essential in encoding and decoding processes.
10. *Chromosome-Parameter* is basically a cascade of substring parameters and contains additional information, namely total chromosome

length, and is used in random generation of cross-sites for crossover operation and overall decoding of vector x.

11. *GA-Parameter* contains information regarding parameters such as crossover and mutation probabilities that can be statically or dynamically changed over generations.

The aforementioned parameter objects (substring and chromosome) are separated from actual objects to prevent multiplicity as many instances of substrings and chromosomes, all having same parameters. Figure 2.4 shows an object model of GA.

In conclusion, the concepts and development of GA in the computational arena has a successful output in every area of science and engineering. Most of the applications that employ GA will perform as a tool for optimization. Basically, when there exist no guidelines to optimize a function of several variables, it works as a random search and try to find an optimal (or 'best or near optimal') solution. Development of concepts of artificial intelligence

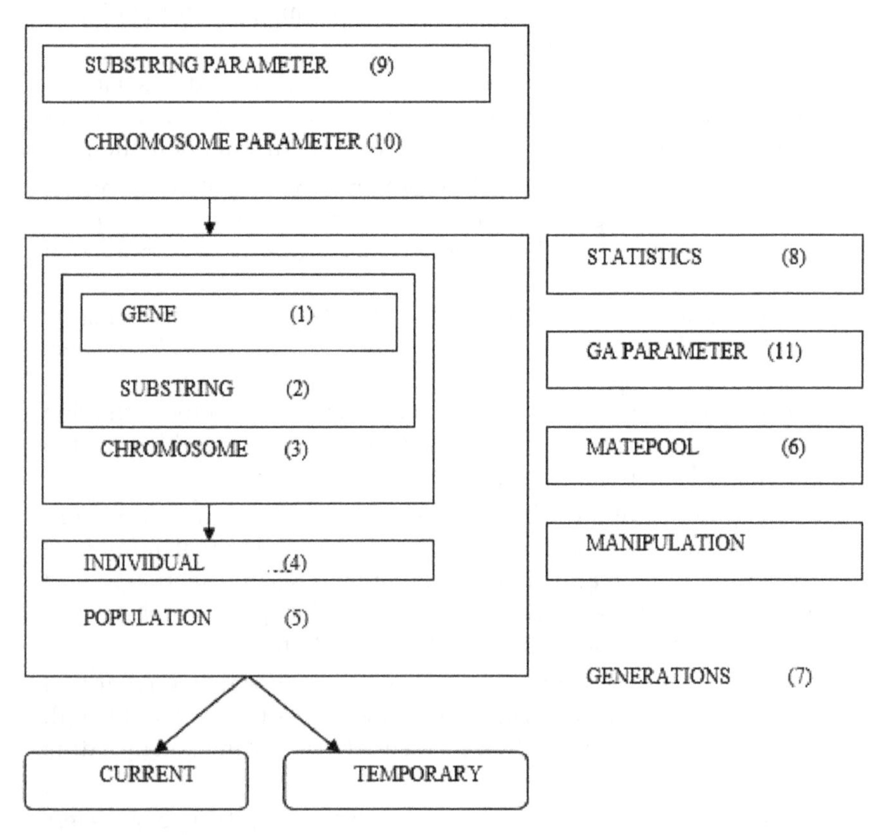

FIGURE 2.4
Object model of genetic algorithms.

(AI) coupled with incomplete information, provides massive applicability of GA in every area of engineering, and management science. However, significant progress is yet to come, and further study is envisaged. Next-generation intelligent machines are likely to employ GA with neuro-fuzzy models to reason and learn from incomplete/missing data and knowledge bases. Such systems will find immense applications in robotics, knowledge acquisition and image understanding systems.

2.2.2 Simulated Annealing

SA is a different subsidiary-free optimization strategy that has lately drawn great consideration for being reasonable for consistency concerning discrete (combinatorial) optimization issues. SA was obtained from physical qualities of turn glasses. Guideline behindhand SA is comparable to what occurs when metals are cool at a controlled rate. Gradually falling temperature enables molecules in liquid metal to arrange them and shape a standard crystalline structure that may have high thickness and low vitality. Besides, if temperature goes down too rapidly, molecules do not have sufficient energy to situate themselves into a standard structure and result in a more nebulous material with higher vitality.

SA is an interconnected worldwide optimization system that crosses search space by testing irregular mutations on a discrete arrangement. A mutation with expanding fitness is constantly acknowledged. A mutation that brings down fitness is acknowledged probabilistically in the light of distinction in fitness and a diminishing temperature parameter. SA has a 'cooling component' (alluded to as 'temperature') that at first enables moves to less-fit arrangements. The impact of cooling on recreation of annealing is that the likelihood of following a negative move is lessened. This at first enables a search to move far from nearby optima in which the search may be caught. While limiting, target work is typically alluded to as a 'cost function'; however, to maximize, it is normally alluded to as a 'fitness function'. In SA, one pinpoints to look for the most minimal vitality rather than the greatest fitness. SA can likewise be utilized inside a standard GA by beginning through a moderately high rate of mutation and diminishing it after some time alongside a specified timetable.

The technique of annealing has been useful to numerous problems in Operational Research and Management Science areas such as cell formation and lot-sizing. It is well known that SA works well for certain types of minimization problems but is not suitable for all types. SA algorithm used in most of areas is very robust in the sense of cooling schedule.

2.2.2.1 Terminologies of SA

Objective function: An objective function $f(.)$ maps an input vector x into a scalar E: $E = f(x)$, where every x is seen as a point in an input space. Assignment of SA is to test input space successfully to discover x that limits E.

Generating function: A generating function $g(.,.)$ determines a likelihood thickness capacity of contrast between the current point and the next point to be gone by.

Acceptance function: After another point x_{new} has been assessed, SA chooses whether to acknowledge or dismiss it in view of estimation of an acceptance function $h(.,.)$. The most commonly applied acceptance function is Boltzmann probability distribution.

$$h(\Delta E, T) = \frac{1}{1 + \exp(\Delta E / (cT))}$$

where $c$is a system-dependent constant, T is the temperature, and ΔE is the energy difference between x_{new} and x, represented by a functional form of $\Delta E = f(x_{new}) - f(x)$.

2.2.2.2 Annealing Schedule

An annealing plan controls how quickly temperature T goes from higher to lower esteems, as an element of time or cycle tallies. The correct understanding of high and low and also the determination of a decent annealing plan need definite issue-particular physical experiences and/or trial and error.

Based on the earlier terminologies, steps to perform SA may be represented in the following manner:

Step 1: Select a starting point x and fix a high starting temperature T. Fix iteration count k to 1.

Step 2: Estimate objective function $E = f(x)$.

Step 3: Choose Δx with probability dogged by generating function $g(\Delta x, T)$. Fixnew point x_{new} equivalent to $(x + \Delta x)$.

Step 4: Compute a new value of objective function $E_{new} = f(x_{new})$.

Step 5: Set x to x_{new} & E to E_{new} with probability calculated by acceptance function $h(\Delta E, T); \Delta E = (E_{new} - E)$.

Step 6: Decrease temperature T conferring to an annealing schedule (typically by just setting T equal to ηT, where η is a constant between 0 and 1).

Step 7: Increment iteration count k. If k touches maximum iteration count, discontinue iteration. Otherwise, go back to step 3.

2.2.2.3 Stepwise Procedure of SA

From an underlying arrangement, the procedure of SA more than once produces a neighbour of current arrangement and exchange to it as per some system with the point of enhancing target work esteem (to expand objective function). Amid this procedure, SA has plausibility to visit more regrettable neighbours, keeping in mind the end goal to escape from 'local' optima. In particular, a parameter called temperature T is utilized to control plausibility of moving to a more awful neighbour arrangement. Algorithm, beginning from a high temperature, over and again diminishes temperature in a key way (called cooling schedule) till the point when temperature is sufficiently lower or some extra stooping criteria is fulfilled. Algorithm acknowledges every single 'great' move and some of the 'terrible' moves as indicated by 'Metropolis probability', characterized by $\exp(-\delta/T)$, where δ is diminished in target work esteem.

Further, for running SA, one of the prime tasks is to state a suitable neighbourhood function so that this generation function assures each feasible solution within the search space. We, then, fix an initial solution, and there are also control parameters that need to be setup (such as initial temperature, cooling ratio and stopping criterion).

In optimization problem, proposed algorithm works as follows:

Define an objective function f and a set of heuristics H.

Define a cooling schedule: starting temperature $t_s > 0$, a temperature reduction function ϕ, a number of iterations for each temperature *temp* and termination criterion.

```
Selectan initial solution s0;
Repeat
    Randomly choose a heuristic h∈H;
    iteration_count = 0;
Repeat
    iteration_count ++ ;
           applying hto s0, get a new solution s1;
δ = f(S1) − f(S0)
if (δ ≥ 0) then s0=s1
else
                produce a random x uniformly in  range (0,1)
                if x<exp(δ/T) then s0=s1;
Until iteration_count =ntemp;
Set 't' = φ(t)
Until stopping criteria = true
End
```

The performance of SA in performing an optimization problem is based on the following criteria:

Initial solution: This was produced uniformly at random contained by a user-defined range.

Initial temperature: This was produced by means of the following procedure: (1) generate specific number of solutions at random;

(2) find solutions with two optimized costs and (3) initial temperature was set at twice difference amongst two optimized costs. In each run, initial temperature is allowed to be adjusted by the user.

Cooling schedule: In the developmental phase of proposed algorithm, we use different types of cooling schedule to find the global optimum. Popular use of this type is as $T = \zeta * T$, where ζ is a constraint that can be attuned by the user. Optimal setting of ζ is extremely problem dependent and robust in nature. General practice to search a near-optimal ζ is to begin with a reasonably smaller value and then progressively turn it to a large value near to 1.0 (say 0.98/0.99), and this trial-and-error process is very monotonous. The second type of cooling schedule that one may use is $T = T/_{1+\theta T}$, where θ is a parameter adjusted by the user and is strictly problem dependent. Again, there is no universally optimal θ that is best for every problems and is strictly problem dependent.

Stopping criterion: Based on the inner loop, this was fixed by the user by the number of repetitions for which temperature T set aside is similar. The outer loop criterion was calculated by the maximum number of iterations set by the user and also the lowest temperature. This implies that the algorithm of SA stops when its temperature is below the lowest value.

2.2.3 Random Search

Random Search investigates the parameter space of an objective function consecutively in an apparently irregular manner to discover the optimal point that minimize or maximize objectives function. Beginning with the most primitive thought, let $f(x)$ be an objective function to be minimized, and x is a point right now under consideration. Random search method is used to find the optimal x by iterating the following four steps:

Step 1: Select a start point x as present point.

Step 2: Add a random vector dx to present point x in parameter space and calculate objective function at a new point at $(x + dx)$.

Step 3: If $f(x + dx) < f(x)$, set current point x equal to $(x + dx)$.

Step 4: Discontinue if the maximum number of function evaluations is obtained otherwise, and then, go back to step 2 to discover a new point.

This is a truly random method in the sense that search directions are purely guided by a random number generator. To improve the earlier method, we follow two observations:

1. If search in a direction results in a higher objective function, opposite direction can often lead to an objective function.
2. Successive successful searches in a certain direction should bias subsequent searching towards this direction, and successive failure in a certain direction should discourage subsequent searching along this direction.

Observation 1 leads to a reverse step in the original method, whereas Observation 2 motivates the use of a bias term as center for random vector. Based on the earlier guidelines, a modified random search method involves the following steps:

Step 1: Select an initial starting point x as present point. Fix initial bias b equal to a zero vector.

Step 2: Add a bias term b and a random vector dx to present point x in input space and calculate the objective function at a new point $(x + b + dx)$

Step 3: If $f(x+b+dx) < f(x)$, set current point x equivalent to $(x+b+dx)$ and bias b equivalent to $(0.2b+0.4dx)$; go to step 6. Otherwise, move to the next step.

Step 4: If $f(x+b-dx) < f(x)$, set present point equivalent to $(x+b-dx)$ and bias b equivalent to $(b-0.4dx)$; go to step 6. Otherwise, move to the next step.

Step 5: Fix bias equal to $0.5b$ and go to step 6.

Step 6: Stop if the maximum number of function assessments is reached. Otherwise, step down to step 2 to discover a new point.

Unlike GAs and SA, random search method is primarily used for continuous optimization problems. It is possible to come up with a random search method for discrete or combinatorial optimizations, but then, preceding observations may no longer be true, and we might lose advantage of modified methods.

2.2.4 Downhill Simplex Search

Downhill Simplex Search is a derivative-free heuristic method for multidimensional function optimization. Idea driving downhill simplex search is basic, and it has a fascinating geometrical understanding. We consider minimization of a component of n factors without any requirements. We begin with an underlying simplex, which is a gathering of $(n+1)$ points in n-dimensional space. Downhill simplex search more than once replaces the point having most elevated function value in a simplex with another point.

A full cycle of downhill simplex search involves the following steps:

Reflection: Define reflection point P^* and its value y^* as

$$P^* = \overline{P} + \alpha\left(\overline{P} - P_h\right),$$

$$y^* = f\left(P^*\right)$$

The reflection coefficient α is a positive constant. Thus, P^* is online joining P_h and $\overline{P}$, on the far side of $\overline{P}$ from P_h. Depending on the value of y^*, we have the following actions:

1. If y^* is in interval 1, go to expansion.
2. If y^* is in interval 2, replace P_h with P^* and finish this cycle.
3. If y^* is in interval 3, replace P_h with P^* and go to contraction.
4. If y^* is in interval 4, go to contraction.

Expansion: Define expansion point P^{**} and its value y^{**} as

$$P^{**} = \overline{P} + \gamma\left(P^* - \overline{P}\right),$$

$$y^{**} = f\left(P^{**}\right)$$

where expansion coefficient γ is greater than unity. If y^{**} is in interval 1, substitute P_h with P^{**} and finish this cycle. Otherwise, replace P_h with the original reflection point P^{**} and finish this cycle.

Contraction: Define contraction point P^{**} and its value y^{**} as

$$P^{**} = \overline{P} + \beta\left(P_h - \overline{P}\right),$$

$$y^{**} = f\left(P^{**}\right)$$

where contraction coefficient β lies between 0 and 1. If y^{**} is in interval 1, 2, or 3, replace P_h with P^{**} and finish this cycle. Otherwise, go to the next step.

Shrinkage: Substitute each P_i with $(P_i + P_l)/2$.

Finish this cycle.

The foregoing cycle is repetitive until a given stopping criterion is met.

In conclusion, we had discussed four most widely held derivative-free optimization methods: namely GAs, SA, random search and downhill search. These techniques rely on latest high-speed computers; they all require a significant amount of computation when compared with derivative-based approaches. However, these derivative-free approaches are more flexible in terms of incorporating intuitive guidelines and forming objective functions.

2.3 Stochastic Optimization – Recent Advancement

2.3.1 Particle Swarm Optimization

It is a populace-based stochastic optimization strategy that is utilized for optimization of ceaseless non-direct capacities. Swarm knowledge can be characterized as a field that covers 'any endeavour to plan algorithms or circulated critical thinking gadgets propelled by accumulation conduct of social creepy crawly settlements and other creature social orders'. Inspiration from angle schools and flying creature used by individuals to enable them to attempt synchronized development, without impact, has been the investigation of enthusiasm for various researchers. There is an overall conviction that social sharing of data amongst people of a populace can give an evolutionary benefit, and there are various cases originating from nature to help; this is center thought behind advancement of particle swarm optimization (PSO).

PSO is a progressive and wide class of Swarm Intelligence technique, and is as of late imagined as a superior analyser that has a few exceedingly alluring characteristics, considering the reality that essential algorithm is extremely clear to comprehend and execute. However, like GAs and evolutionary algorithms, it needs less computational accounting and for the most part less lines of code. PSO started investigations of synchronous fledgling rushing and fish tutoring, when examination understood that their reenactment algorithms had an improving trademark. Eberhart and Kennedy initially offered PSO as a recreation of social conduct, and it was at first presented as an optimization strategy in 1995.

All things considered, flying creatures do not have a clue (for example) where sustenance is; however, in the event that one puts out a winged creature feeder, he/she will see that inside hours an extraordinary number of flying creatures will find that it's there, despite the fact that they had no past learning about it. This looks like rush progression empowers individuals from run to gain by each other's learning. Operators can, along these lines, be acclimatized to arrangement seekers that socially share learning while they fly over an answer space. Every operator that has discovered anything great leads its neighbours towards it, so that in the end they can arrive on the 'best' arrangement in field.

Particle Swarm upgrades a target work embraced by a populace-based search. Populace comprises of potential arrangements, termed particles that are a representation of flying creatures in herds. These particles are haphazardly introduced and unreservedly fly crosswise over a multi-dimensional search space. Amid flight, every particle refreshes its individual speed and position in light of finest involvement of its own and whole populace. Refreshing strategy drives particle swarm to push to an area with higher target work esteem, and in the long run, all particles will accumulate around a point with the most noteworthy target esteem.

Stepwise procedure for development of PSO is as follows:

Step 1: *Initialization* – At this time, velocity and position of all particles are arbitrarily set to within pre-defined ranges.

Step 2: *Velocity updating* – At every iteration, velocity of all particles are refreshed by

$$\vec{V}_i = w\vec{V}_i + c_1R_1\left(\vec{p}_{i,\text{best}} - \vec{p}_i\right) + c_2R_2\left(\vec{g}_{i,\text{best}} - \vec{p}_i\right)$$

where $\vec{p}_i$ and $\vec{V}_i$ are the position and velocity of particle i individually, $\vec{p}_{i,\text{best}}$ and $\vec{g}_{i,\text{best}}$ are the positions with the 'best' objective function discovered so far by particle i and whole populace separately, w is a parameter governing flying dynamics, R_1 and R_2 are arbitrary variable in go $[0,1]$ and c_1 and c_2 are the factors controlling related weighting of comparing terms. Incorporation of arbitrary variable enriches PSO with the capacity of stochastic search. Weighting factors c_1 and c_2 deal unavoidable exchange off between investigation and exploitation. In the wake of refreshing, $\vec{V}_i$ ought to be checked and secured inside a pre-determined range to stay away from violent irregular walking.

Step 3: *Position updating* – Assuming a unit time interim amid progressive emphases, places of all particles are refreshed by

$$\vec{p}_1 = \vec{p}_i + \vec{V}_i$$

After updating, $\vec{p}_i$ must be checked and restricted to a permitted range.

Step 4: *Memory updating* – Update $\vec{p}_{i,\text{best}}$ & $\vec{g}_{i,\text{best}}$ while the conditions are encountered.

$$\vec{p}_{i,\text{best}} = \vec{p}_i \text{ if } f\left(\vec{p}_i\right) > f\left(\vec{p}_{i,\text{best}}\right)$$

$$\vec{g}_{i,\text{best}} = \vec{g}_i \text{ if } f\left(\vec{g}_i\right) > f\left(\vec{g}_{i,\text{best}}\right)$$

where $f(\vec{x})$ is the objective function which is to be maximized.

Step 5: *Termination checking* – Algorithm repeats steps 2–4 till firm end conditions are met; for example, a pre-characterized number of emphases or an inability to progress for a specific number of cycles. Once ended, algorithm reports estimations of $\vec{g}_{\text{best}}$ and $f\left(\vec{g}_{\text{best}}\right)$ as its solution.

2.3.1.1 Discussion of PSO over GA in Soft Computing Platform

Quality of GAs is parallel nature of infiltrating/search nature. A GA executes a capable type of slope climbing that jelly various arrangements, destroys unpromising arrangements and gives sensible arrangements. Genetic administrators utilized are fundamental to the accomplishment of search. Every GA needs some type of recombination, as this permits formation of new arrangements that take, by temperance of their parent's prosperity, a higher likelihood of showing a decent execution. By and large, crossover is the primary genetic administrator, though mutation is utilized significantly less often as possible. Crossover endeavours to protect valuable parts of competitor arrangements and to wipe out bothersome segments, while irregular nature of mutation is most likely more inclined to corrupt a solid applicant arrangement than to enhance it. Alternative wellspring of algorithm's energy is understood parallelism inborn in evolutionary metaphor; by limiting generation of weak competitors, GAs take out that arrangement as well as the greater part of its descendants. This inclines to make an algorithm liable to unite to astounding arrangements inside a couple of ages.

Particle Swarm imparts numerous similitudes to evolutionary calculation strategies by and large and GAs specifically. Every one of these procedures starts with a gathering of a haphazardly created populace; all use a fitness incentive to assess populace. They all refresh populace and search for ideal with irregular procedures. In any case, primary contrast between PSO approaches contrasted, and GA in soft computing stage is that PSO does not have genetic administrators, for example, crossover and mutation. Particles refresh themselves with inside speed; they additionally have memory that is significant to algorithm. In contrasted and EC class of algorithms, data-sharing component in PSO is considerably unique. In EC methods, chromosomes share data with one other, and in this manner, the entire populace travels like one gathering towards an ideal region. In PSO, just the 'best' particle provides out data to others. It is a one-way data-sharing component, and evolution searches for best arrangement. In contrasted and ECs, all particles have a tendency to meet the best arrangement immediately even in nearby form. Once more, in contrast to GAs, the benefits of PSO are that PSO is relatively simple to execute and there are a couple of parameters to alter. Be that as it may, GA can be seen as one of the best class of optimization procedure on account of its fluctuation of use in various building and administration science issues.

2.3.2 Ant Colony Optimization

Ant colony optimization (ACO) is a novel and exceptionally encouraging research field found at the intersection between Artificial life and Operational Research. This (otherwise called Ant algorithms) is a worldview for planning meta-heuristic algorithms for combinatorial optimization created by Dorigo and Gamberdella. Ant algorithms are organically roused from conduct of settlements of genuine ants, and specifically, how they scavenge for sustenance. Basic characteristic of ACO algorithms is a blend of the earlier data about the structure of promising arrangement with posteriori data about structure of beforehand acquired great arrangement. Meta-heuristic algorithms are the ones which, keeping in mind the end goal to escape from nearby optima, drive few fundamental heuristic: moreover, a productive beginning from an invalid arrangement and adding components to fabricate a decent entire one, or a neighbourhood search heuristic beginning from a total arrangement and iteratively changing some of its components with a specific end goal to accomplish a superior one. Meta-heuristic part allows low-level heuristic to get arrangements superior to anything those it could have accomplished alone, regardless of whether iterated. Typically, the controlling component is accomplished either by obliging or by randomizing a set of nearby neighbour answers for consideration in neighbourhood search.

Advancement of ACO applications is a notable symmetric and topsy-turvy traveling salesman problem (TSP). An immediate expansion of TSP, the first issue of Ant framework, was connected to vehicle routing problems (VRP), where an arrangement of vehicles positioned at a terminal needs to serve an arrangement of clients before coming back to station, and the target is to limit all the while the number of vehicles being utilized and add up to separate passed by vehicles. Quadratic Assignment Problem is another sort of operational research/administration science issue of doling out n offices to n areas in this way, to the point that task cost is limited. Here, cost is characterized by quadratic capacity, and this issue can be tackled effectively by utilizing a system of ACO.

VRP/dispersion of coordinations concern transport of products from terminals and clients by methods for an armada of vehicles. VRP can be instantiated to some certifiable areas: basic cases are perishable gliding of drain item, mail conveyance, circulation of oil items, distribute up and conveyance framework and so on. Explaining a VRP intends to discover 'best course' to benefit all clients utilizing an armada of vehicles. Basic target is minimization of transportation costs as a component of voyaged remove or of voyaged time. Settled expenses might be considered, and in this manner, the number of vehicles is additionally limited. Another goal is to feature vehicle effectiveness communicated as a level of load limit. Arrangement must guarantee benefit operation parameter that all clients are being served, keeping every single operational limitation, minimum to transportation cost.

Along these lines, VRP can be figured as a progressed numerical programming issue, with a goal of work and an arrangement of limitations. Misusing the qualities of numerical demonstration, ACO algorithm can be produced and changed to discover ideal or close ideal answer for proficient administration of a conveyance procedure in store network. Lately, this progressed stochastic optimization strategy is differentiated in various fields such as Job-shop booking, Vehicle directing, Sequential requesting and Graph shading. Fruitful utilization of ACO has raised genuine regard for some of the organizations for taking care of genuine administrative issues. Euro Bios first utilizes it to create algorithms in light of ACO meta-heuristic; they have created it for taking care of various scheduling issues (Continuous two-step flow shop issue with limited repositories). Another promising utilization of ACO has been created by Ant Optima by the same Bios Group of optimization analyst for tackling diverse kinds of Vehicle Routing issues in Logistics and Distribution fields.

Travelling Salesman Problem is another famous issue in writing and has pulled in a lot of research intrigue. In the first place, ACO algorithm, called Ant System, and in addition a large number of ACO algorithms proposed henceforth, was first tried on TSP, and it is a standard proving ground for new algorithmic thoughts. Concerning applications, utilization of ACO for arrangement of dynamic, multi-objective, stochastic, consistent and blended variable optimization issues is a hotly debated issue, and also the production of parallel usage equipped for exploiting new accessible parallel hardware. Ebb and flow research includes dynamic issues and powerful steering of information bundles through web utilization in a most brief conceivable way. A few researchers are at present occupied with the use of ant algorithms to picture handling, all the more particularly to picture division and bunching algorithms. There is an awesome desire in the field of AI that new systems will fill a hole left by conventional image-based AI, where learning and managing new circumstances is dependably a gigantic test. Ant algorithms are one of such encouraging methodologies, and there is still much work left to be finished.

2.3.2.1 ACO – How It Works in Any Optimization Issue?

The conduct of a solitary ant, honeybee and termite are excessively basic, yet their aggregate and social conduct is of significance in nature. Contrasting and insight, the mental aptitude of human is for the most part extensive in contrast to the individual ant or honeybee. However, when ant's work is meant in an 'aggregate manner', genuine power can be assessed by their pilgrim mind. A few of ant's conduct can be actualized in a computational situation, keeping in mind the end goal to take care of a specific issue or to determine a new conduct such as ant settlement conduct. Two essential inquiries on which we are intrigued about the ant's conduct are as follows:

- How real ants execute the arrangement of searching for nourishment?
- How they interface to transport it from the discovered area to their home, in the least conceivable time?

ACO algorithms were fundamentally roused by the perception of genuine ant states. Ants are social creepy crawlies that live in states and whose conduct is thought more to survival of colony all in all as opposed to a solitary individual part. Social bugs have caught the consideration of numerous researchers because of high structuration level of their settlements. An important and intriguing conduct of ant states is their 'searching conduct' and specifically how ants can discover the most limited ways between sustenance sources and their home.

Social ants, while strolling from sustenance sources to settle and the other way around ants, store on ground a substance called 'pheromone', shaping thusly a pheromone trail. Ants may notice pheromone and, while picking their direction, they have a tendency to pick in 'likelihood ways' set apart by solid pheromone fixations. Pheromone trail enables ants to search their way back to nourishment source/settle. It has been demonstrated tentatively that this pheromone trail following conduct can give rise once utilized by a colony of ants to development of most brief negligible ways. Again, when more ways are accessible from home to a nourishment source, a colony of ants might have the capacity to abuse pheromone trails left by singular ants to find most limited negligible ways from home to sustenance source and back.

In taking care of an optimization issue by ACO, a colony of simulated ants iteratively develops answers for issue under thought by utilizing manufactured pheromone trails and heuristic data. Pheromone trails are changed by ants amid algorithm execution with a specific end goal to store data about 'great' arrangements.

2.3.2.2 ACO – Sequential Steps

ACO is a prototype for designing meta-heuristic algorithms for combinatorial optimization problems. It shows a state-of-art performance for solving several complex application problems. General guideline to tackle any optimization problem by ACO may be summarized as follows:

Step 1: Transform problem in terms of set of components and transitions; in which platform ants may lay/build up solutions.

Step 2: Represent pheromone trails effectively to problems we are to deal with.

Step 3: Sequential heuristic preference for every decision that a colonial ant has to decide while moving towards optimal solution.

Step 4: Successful implementation of an efficient LS algorithm that pursuits problem on hand.

Step 5: Perform algorithm.

Step 6: Fine-tuning of all parameters of ACO algorithm.

A decent initial point for parameter tuning may get even better result to a variety of comparative problems. These aforementioned steps are merely basic operation – management scientists may enhance or change some choices and also revise or modify according to complexity and behavior of problem.

2.3.2.3 ACO – Meta-Heuristics Approach

Ants are exceptionally basic animals with an extremely low subjective limit. Aggregate assemblage of colony is fit for dealing with different complex undertakings as home development, upkeep and sustenance gathering. It is found by ethnologists that leading ants build most brief way from their colony to encouraging source through utilization of pheromone trails. The process is imitated in ACO over utilization of an arrangement of straightforward specialists (i.e., simulated ants) that were dispensed with computational assets and they abuse stigmergic correspondence – a type of aberrant correspondence interceded by condition to discover answers for issues close by. Amid development, ants for the most part drop a certain measure of pheromone – a compound substance utilized by ants to impart and trade data about which course to take after – on their picked way accordingly checking it with a trail of this substance. The approaching ant detects pheromone in various ways and probabilistically chooses a way as per likelihood that is straightforwardly corresponding to measure the pheromone on it.

Throughout, number of counter (NC) cycle for kth ant on node i, selection probability of next node j to trail, is given by

$$\phi_{ij}^{k} = \frac{\left[\tau_{ij}(\mathrm{NC})\right]^{\alpha}\left[\eta_{ij}(\mathrm{NC})\right]^{\beta}}{\sum_{k \in \mathrm{feas}_k}\left[\tau_{ij}(\mathrm{NC})\right]^{\alpha}\left[\eta_{ij}(\mathrm{NC})\right]^{\beta}}, j \in \mathrm{feas}_k$$

$$= 0; \text{otherwise}$$

where η_{ij} is a heuristic value called visibility on edge (i, j)

τ_{ij} is the amount of trail laid on edge (i, j)

Step 1: Initialize control parameters.

Control parameters of ACO that have initialized as total number of ants (N = total number of operations), relative importance of trail $(\alpha = 1.0)$, relative importance of visibility $(\beta = 1.0)$ and pheromone evaporation rate $(\rho = 0.50)$

Step 2: Initialization

a. Problem representation using a weighted directed graph
b. Randomly distribute ants on node
c. Set $t = 0$ (time counter)
d. Set NC = *0* (NC is number of counter)
e. Set $\mathrm{tabu}^k = 0$ (tabu represents list of nodes traversed by ant k)
f. Select increase in trail level equal to zero.

If NC >NC_max, go to step 8; otherwise, proceed.

Step 3: Condition

If $k > k\_\max$, go to step 7; otherwise, proceed ($k\_\max$ is the number of ants)

Step 4: Condition

If tabu list of ant k is full $\left(\text{tabu}^k \geq \text{tabu}^k\_\max\right)$, go to step 6; otherwise, proceed.

Step 5: Node selection

a. Generate a random number $p\left(0 \leq p \leq 1\right)$
b. If $p \geq p_0$, proceed; otherwise, go to step 5(d).
c. Compare probability of possible outgoing nodes.
d. Choose a particular node (job) having with highest probability.
e. Add node to tabu^k and remove it from further consideration.

Step 6: Condition

$k = k + 1$, go to step 3.

Step 7: Updating

a. Find $p_{\text{iter}}^{+} < p_{\text{best}}^{+}$, then $p_{\text{iter}}^{+} = p_{\text{best}}^{+}$ (p_{iter}^{+} is the best objective function value).
b. Update pheromone: $\tau_{\text{il}}(t)$.
c. Empty all tabu lists.
d. $\text{NC} = \text{NC} + 1$ (increase counter number)
e. $p_0 = \log(\text{NC})/\log_e(N\_\max)$, go to step 2.

Step 8: Output

p_{best}^{+} (the best objective function value)

End

The earlier stated program terminates if at least one of the following three conditions are satisfied: algorithm has discovered an answer inside a predefined distance from a lower bound on optimal arrangement quality, or time complexity will show reasonably good results, or maximum CPU time has been consumed or efficiency of algorithm comes in a stagnation position.

In conclusion, biologically inspired computing is recently given importance for its immense parallelism and simplicity in computation for handling complex problems. ACO is a population-based advanced metaheuristic approach that can be used to evaluate approximate solutions/sub-optimal solutions to difficult optimization problems. Fundamental attribute of ACO algorithms is blend of the earlier data about structure of a promising solution with a posteriori data about structure of beforehand acquired great solution.

2.4 Hybridization

Applying GA, the main drawback that suffers this traditional approach are slow convergence rate and premature convergence. To overcome and improve upon such defects, a new concept is developed by recent researchers, popularly known as *hybridization* within GA or hybrid GA. The basic concepts of using hybrid GA are as follows: combine simple GA with efficient LS algorithm. The purpose of using these concepts is to split up/distribute optimization task into two parts: simple GA first performs search operation followed by refinement and is done by LS algorithm. Both algorithms can run in parallel, say after *'n'* iteration of LS local optimal solution is injected into current generation. An LS method locates the 'local minima' that complements simple GA to achieve/move towards 'global minima'.

An assortment of procedures has been crossbred to embody 'best of both' strategies in true building and administration issues. Strategy of hybridization of learning and worldwide GA is memetic algorithm (MA). MA is inspired by Dawkins documentation of an image. An image is a unit of data that duplicates itself as individuals trade thoughts. Mama ties usefulness of GA with a few heuristic search strategies such asslope climbing, SA and Tabu Search. Two well-known methods for hybridization rely upon the ideas of 'Baldwin impact' and 'Lamarckism'. As per Baldwinian search procedure, neighbourhood optimization can collaborate and permit nearby search space to change fitness of an individual without changing its genotype itself. Fundamental disadvantage of Baldwinism search procedure is that it is moderate. As indicated by Lamarckism, qualities gained by an individual amid its lifetime may end up noticeably heritable attributes. As per this approach, fitness and genotype of people are changed amid nearby optimization stage. The vast majority of MA depends on Lamarckism approach of hybridization. Proposed MA consolidates slope climbing neighbourhood search after determination process with a specific end goal to build abuse. In this approach, individuals chose utilizing roulette wheel determination has been utilized as a starting point to complete slope-climbing search. Here, every individual is enhanced utilizing slope, moving before going to propagation stage. Researchers exhibited the impact of a nearby search strategy in diminishment of populace measure. By and large, mutation and crossover administrators deliver infeasible answers for a profoundly obliged issue. To keep away from age of infeasible arrangements, numerous methods have been proposed like incomplete coordinated crossover (Partial Matched Crossover) for use all together-based issues. To take care of exceptionally compelled timetabling issue, a heuristic crossover administrator was presented with coordinate portrayal of timetable, so essential limitations are never abused for taking care of voyaging salesperson issue-adjusted crossover (Modified Crossover).

Conceivable explanations behind hybridization can be expressed in the following way: to enhance the nature of arrangements acquired by evolutionary algorithm, to enhance the execution of evolutionary algorithm and to join evolutionary algorithm as a feature of a bigger framework. There are a number of approaches to fuse different strategies from introduction of populace to the age of offspring. Populace might be instated by fuse of earlier learning, heuristics, neighbourhood search and so forth. Nearby search methods might be connected amid instatement or amid age of offspring. Fundamental process includes hybridization to conquer confinement of individual strategy; diverse learning and adjustment methods are coordinated through hybridization or a combination of these strategies. Because of the absence of a typical structure, it remains regularly hard to think about and assess execution of different half-and-half systems thoughtfully and nearly. There are a few approaches to incorporate straightforward GA with other meta-heuristics. Some of them might be coordinated in line of: arrangements of beginning populace of GA might be made by issue-particular heuristics, a few or all arrangements acquired by evolutionary algorithms in soft computing methodology might be enhanced by nearby search and arrangements might be spoken to in a roundabout way and a disentangling algorithm maps any genotype to a comparing phenotypic arrangement. In this mapping, decoder can utilize issue particular qualities and apply heuristics and so on.

Utilization of hybridization might be appeared by celebrated work as done by Jeong and his research group. Jeong et al. (2005) recommended a cross-breed approach with a GA and a reenactment strategy. GA is utilized to advance timetables, and recreation is utilized for minimization of the most extreme culmination time for last occupation with settled calendars got from GA display. Principal ventures of approach are creation plans are produced utilizing GA, run a reproduction display in view of GA-produced generation plans, acquire possible reenactment finish time, settle on fitting outcome, which yields required esteems, change imperatives in GA utilizing current recreation culmination time and go to generation plan as indicated earlier and at last decide creation booking that is thought to be a sensible ideal arrangement.

Section II

Various Case Studies Comprising Industrial Problems and Their Optimal Solutions Using Different Techniques

3

Single-Objective Optimization Using Response Surface Methodology

3.1 Introduction

The material removal rate (MRR) in an electro-discharge machining (EDM) process is taken as the machining response influenced by a number of machining parameters such as spark off-time, spark on-time, spark current and gap voltage. For the purpose of experiment, an EN 19 tool steel workpiece is employed, where response surface methodology (RSM) is utilized for finding the characteristics of machining responses with varying machining parameters. Empirical equations are evaluated to correlate the data responses obtained from the experiments with the varying process parameters. The parameters having critical influence on the MRR responses are taken into consideration, and the variations are studied thoroughly.

3.2 Experimentation, Result and Discussion

Experiments are performed by varying input parameters (gap voltage, spark current, spark on-time, spark off-time) and obtaining the output responses. The responses in terms of surface roughness values and MRR are studied. The number of experiments (based on the central composite design [CCD]) conducted is 31, which is presented in Table 3.1. The result of the experiment is used to construct empirical equations using RSM. Minitab software is used for developing a second-order response model.

Empirical equation formulation involves a thorough study of the responses (MRR in this case) and implication of analysis of variance (ANOVA) in subsequent steps to draw a feasible range of variations in response (with change in parameters as mentioned earlier) with the introduction of an F factor, which represents a confidence level of the parameter. F factor is taken for a confidence level of 95%, and the variations having more than 0.05 F value are

TABLE 3.1

Experimental Results

Run Order	A	B	C	D	MRR (g/min)
1	–1	–1	1	1	0.21
2	1	–1	–1	–1	0.13
3	–1	–1	–1	1	0.10
4	0	0	0	0	0.23
5	0	0	0	–2	0.39
6	0	2	0	0	0.33
7	1	–1	1	–1	0.32
8	–1	–1	–1	–1	0.14
9	1	–1	–1	1	0.11
10	0	0	0	0	0.23
11	–1	1	1	1	0.34
12	1	1	–1	–1	0.22
13	1	1	1	1	0.33
14	–1	1	–1	–1	0.22
15	0	0	0	0	0.23
16	0	0	0	0	0.23
17	2	0	0	0	0.24
18	0	0	0	2	0.23
19	0	0	2	0	0.49
20	0	–2	0	0	0.12
21	1	1	–1	1	0.14
22	0	0	–2	0	0.09
23	0	0	0	0	0.23
24	1	1	1	–1	0.49
25	0	0	0	0	0.23
26	–1	1	–1	1	0.15
27	–2	0	0	0	0.23
28	0	0	0	0	0.23
29	–1	–1	1	–1	0.31
30	1	–1	1	1	0.22
31	–1	1	1	–1	0.49

considered insignificant, while the response variations below this value are taken into consideration for constructing relations between responses and variables. The non-linear nature of EDM does not allow linear operations to function due to which a quadratic model was preferred to be applied. While observing the adequacy test by ANOVA, the linear terms $T_{\text{off}}, T_{\text{on}}, I_p, V$ interactive term T_{off} with I_p, I_p with V and square terms V^2 were found to be significant for the formulation of an effective model. Significant levels are depicted in 3.2. The quadratic model was observed to be a significant

TABLE 3.2

ANOVA Table (before Elimination)

Factors	Coefficient	SE Coefficient	*T*	*p*
Constant	0.228	0.005	43.766	0
A	0.001	0.003	0.235	0.807
B	0.053	0.003	20.417	0
C	0.096	0.003	36.016	0
D	−0.043	0.003	−14.316	0
A*A	−0.002	0.003	−0.802	0.436
B*B	−0.003	0.003	−1.293	0.202
C*C	0.013	0.003	6.025	0
D*D	0.018	0.003	6.983	0
A*B	−0.002	0.003	−0.587	0.534
A*C	0.003	0.003	0.619	0.466
A*D	0.000	0.003	0.086	0.929
B*C	0.021	0.003	7.002	0
B*D	−0.013	0.003	−4.013	0.001
C*D	−0.018	0.003	−6.115	0

Notes: $S = 0.013$; $P_{RESS} = 0.017$.
R-Sq, R-Sq (actual) and R-Sq (truncated) are 99.21, 95.46 and 98.52, respectively.

approach statistically for carrying out analysis in MRR, and insignificant variations (p values >0.05) are ignored (Table 3.2).

The reduced model (after eliminating the insignificant variations) in Table 3.3 gives the R^2 values and adjusted R^2 values to be 99.06% and 98.65%, respectively, which indicates that the model is significant for p value less than 0.05. The insignificant variations are eliminated, and a regression equation

TABLE 3.3

ANOVA Table after Backward Elimination

Factors	Coefficient	SE Coefficient	*T*	*p*
Constant	0.228	0.005	43.766	0
A	0.001	0.003	0.260	0.798
B	0.053	0.003	20.293	0.000
C	0.096	0.003	36.675	0.000
D	−0.043	0.003	−16.439	0.000
C*C	0.013	0.002	5.625	0.000
D*D	0.018	0.002	7.691	0.000
B*C	0.021	0.003	6.638	0.000
B*D	−0.013	0.003	−4.048	0.001
C*D	−0.018	0.003	−5.685	0.000

Notes: $S = 0.013$; $P_{RESS} = 0.012$.
R-Sq, R-Sq (actual) and R-Sq (truncated) are 99.06, 96.64 and 98.65, respectively.

is formed, which exhibits the relationship between MRR and the variable parameters given as follows:

$$\text{MRR} = 0.223 + 0.001 * T_{\text{on}} - 0.043 * V + 0.053 * T_{\text{off}} + 0.096 * I_p$$
$$- 0.013 * T_{\text{off}}^2 + 0.018 * V^2 + 0.013 * T_{\text{on}}^2 - 0.0183 * V * I_p + 0.021 * I_p * T_{\text{off}} \quad (3.1)$$

The obtained final model is passed through an F test, the values of which justifies the properness of the model. ANOVA tests were performed to evaluate and ascertain the fitness of the developed model. The tests are done to get the best fit of the model for the analysis of data. The ANOVA table is given in Table 3.4.

The variation in residuals during analysis of MRR is given in Figure 3.1. The residual generally denotes the error difference between experimental and predictive values. By observing Figure 3.1, it is seen that the residuals lie close to the straight line and follows a normal distribution error pattern which is the very expected distribution. The residuals are compared with the fitted model in Figure 3.2, which conforms to the abovementioned normal distribution of errors. Since no irregular pattern in variations in predicted and experimental values have been observed, the model is preferred the best fit for prediction and proves to be significant.

Figure 3.3 presents a plot between the MRR mean values and the process parameters discussed earlier. It is observed that the spark off-time and the spark current exhibit the highest inclination and are therefore considered as significant, while the spark on-time and gap voltage almost exhibit a horizontal variation indicating insignificant parameters.

Three-dimensional response plot has been formed out of the relations given by the regression equation between the input parameters (spark off-time, spark on-time, spark current and gap voltage) and the responses (as in MRR in this case). Figure 3.4 plots the variation of MRR with the spark on-time and the spark current, taking gap voltage and spark off-time to

TABLE 3.4

Analysis of Variance for MRR

Source	DF	Seq SS	Adj SS	Adj MS	*F*	*p*
Regression	9	0.364	0.364	0.040	244.870	0
Linear	4	0.334	0.334	0.084	506.800	0
Square	2	0.014	0.014	0.007	41.930	0
Interaction	3	0.015	0.015	0.005	30.920	0
Residual error	21	0.003	0.003	0.000		
Lackoffit	15	0.003	0.003	0.000		
Pure error	6	0.000	0.000	0.000		
Total	30	0.367				

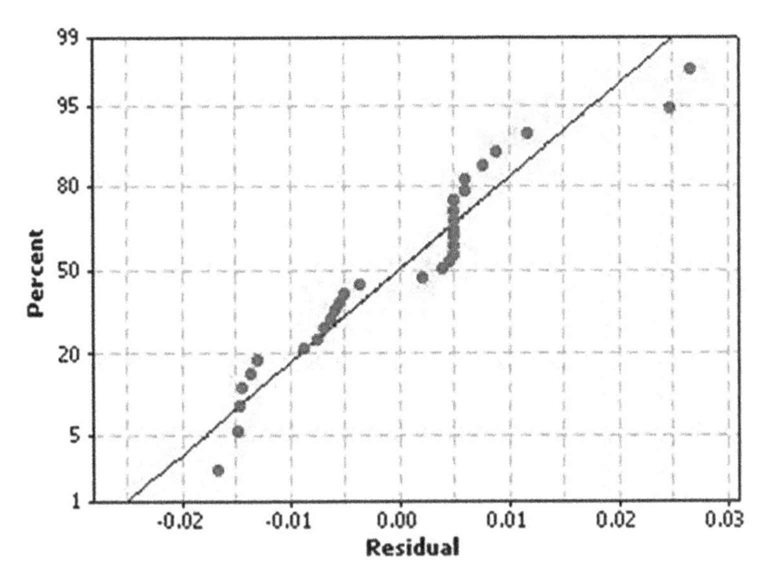

FIGURE 3.1
Residuals for MRR-normal probability plot.

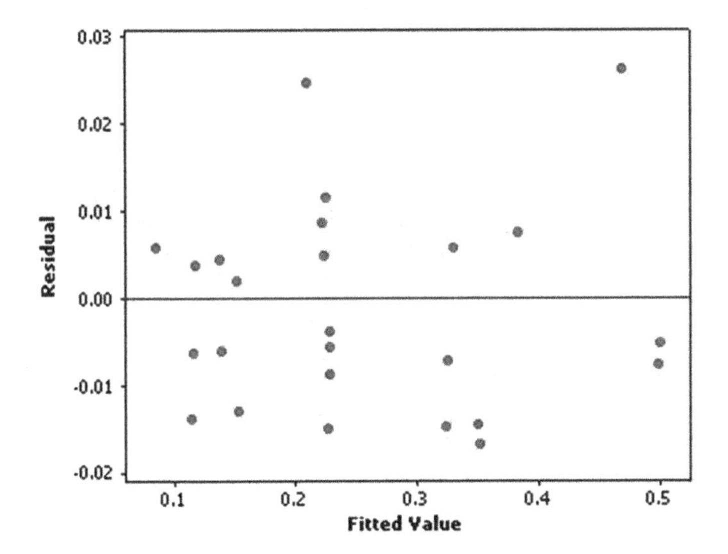

FIGURE 3.2
Variation of plot of residuals and fits for MRR.

remain constant. The slope of the inclination was found to be the greatest towards the higher part of the spark current and the spark on-time. This is in conformation with the fact that at higher spark currents due to high spark energy, the material erosion rate of the workpiece is more at a low spark current. In Figure 3.5, the contour plot showing variations in MRR due

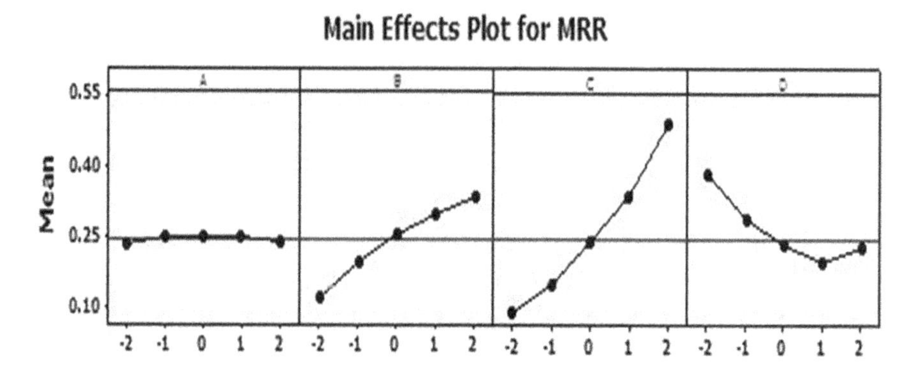

FIGURE 3.3
Main effects plot.

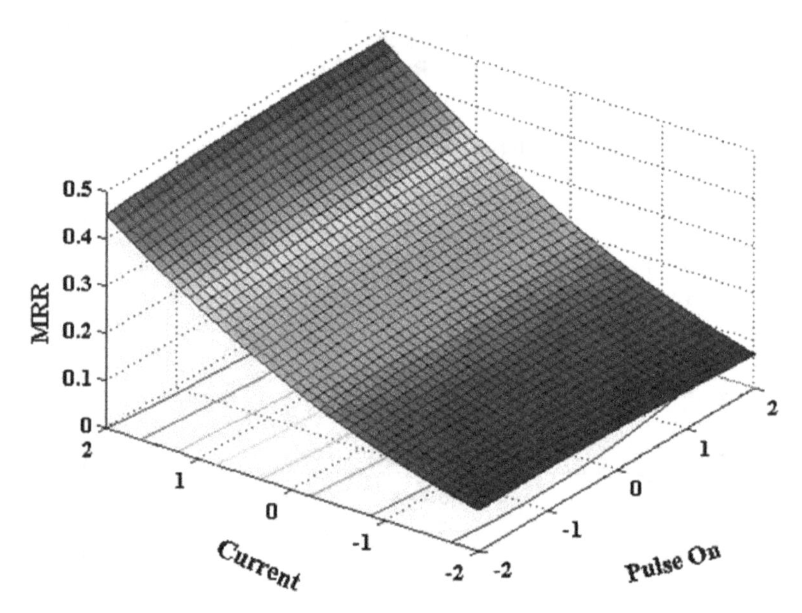

FIGURE 3.4
MRR vs. I_p and T_{on}.

to spark current and spark off-time has been depicted. It is observed that MRR increases with increase in spark current but performs opposite with increase in spark off-time. This is in conformation with the fact that, during the spark off-time period, the spark seizes to function, and there will not be any melting of the workpiece. The heat is evolved out in the form of undesirable loss and the workpiece gets cooled. During the next phase of spark, the material melted is less and the MRR is thus reduced. In Figure 3.6, the variations in MRR with spark current and gap voltage are depicted. It is observed

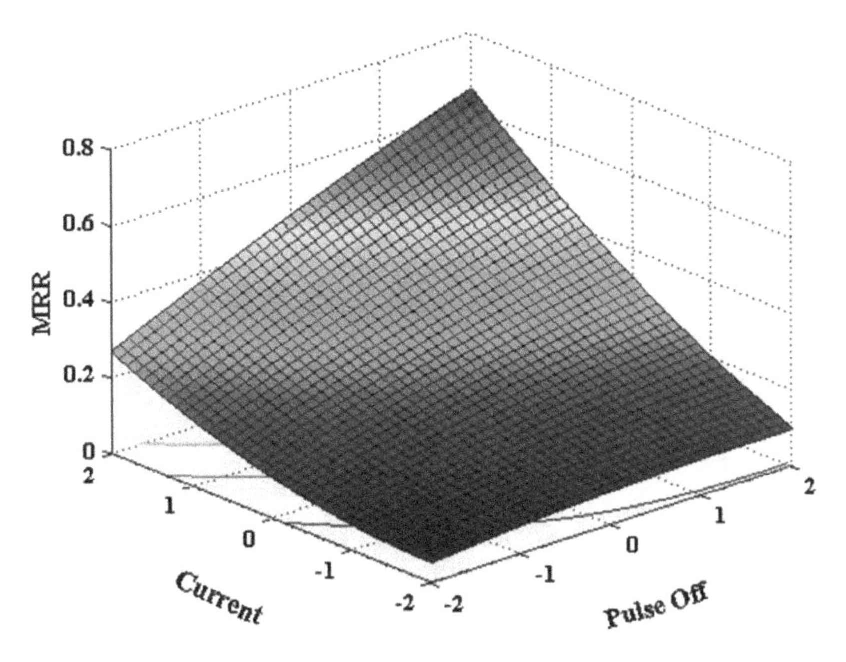

FIGURE 3.5
MRR vs. I_p and T_{off}.

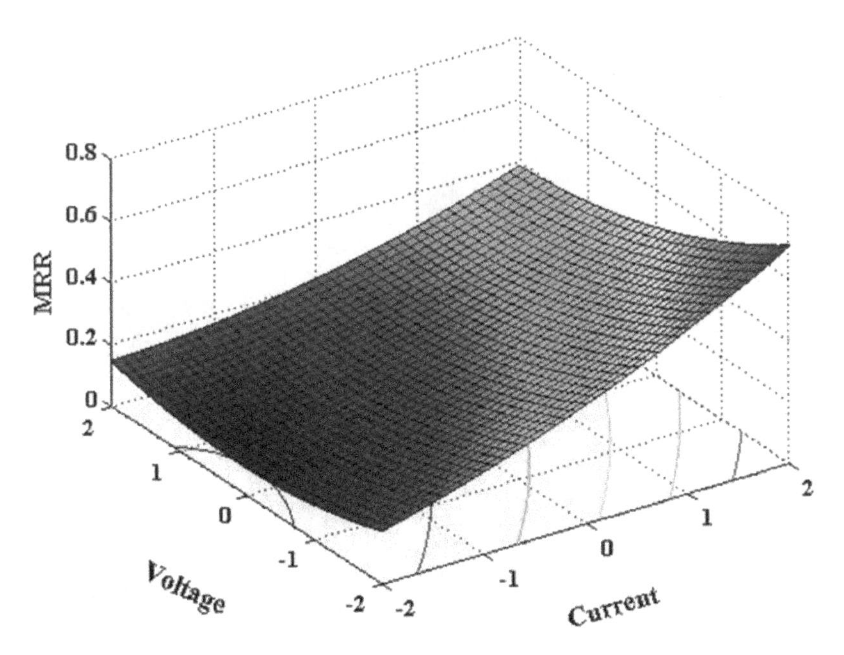

FIGURE 3.6
MRR vs. V and I_p.

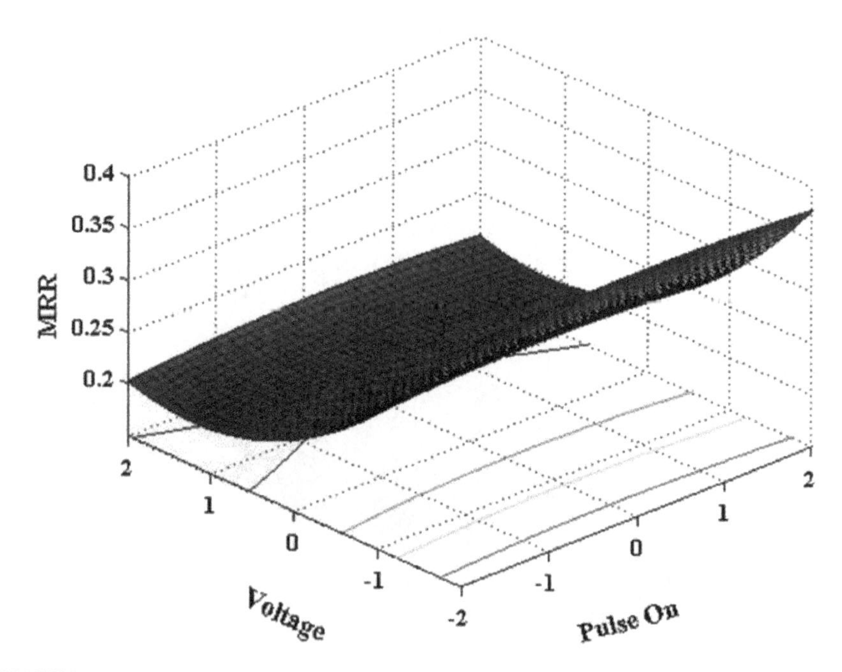

FIGURE 3.7
MRR vs. V and T_{on}.

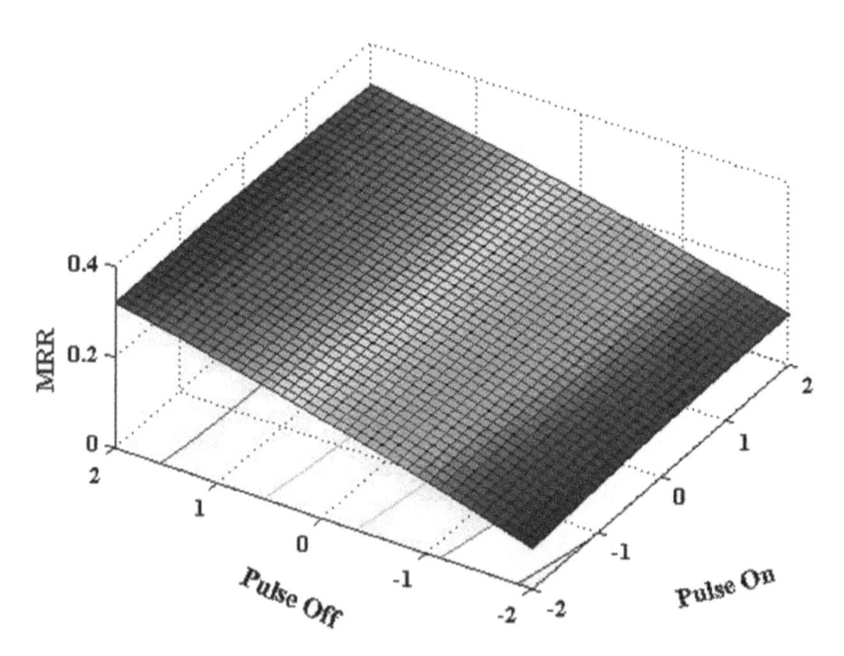

FIGURE 3.8
MRR vs. T_{off} and T_{on}.

that MRR increases rapidly with an increase in spark current, but there is no much effect in MRR with an increase in gap voltage. Figure 3.7 depicts the effect of spark on-time and gap voltage on MRR. It is seen that with an increase in gap voltage, the spark current increases proportionally, which leads to higher MRR along the spark on-time. Lastly, similar response surface plots have also been created considering the effect of spark on-time and spark off-time in Figure 3.8 and spark off-time and gap voltage in Figure 3.9 on MRR.

The final check for the developed model is done by conforming with the experimental result, employing the midterm values of the input parameters. The expected MRR is found to have an error in the experimental MRR which is small (2%) (Table 3.5).

Finally, optimum process parameter values are obtained by implementing the surface response for maximum MRR. The process parameters are found to have the values: $T_{on} = 300$ μs, $T_{off} = 1{,}700$ μs, $I_p = 12$ A and $V = 60$ V.

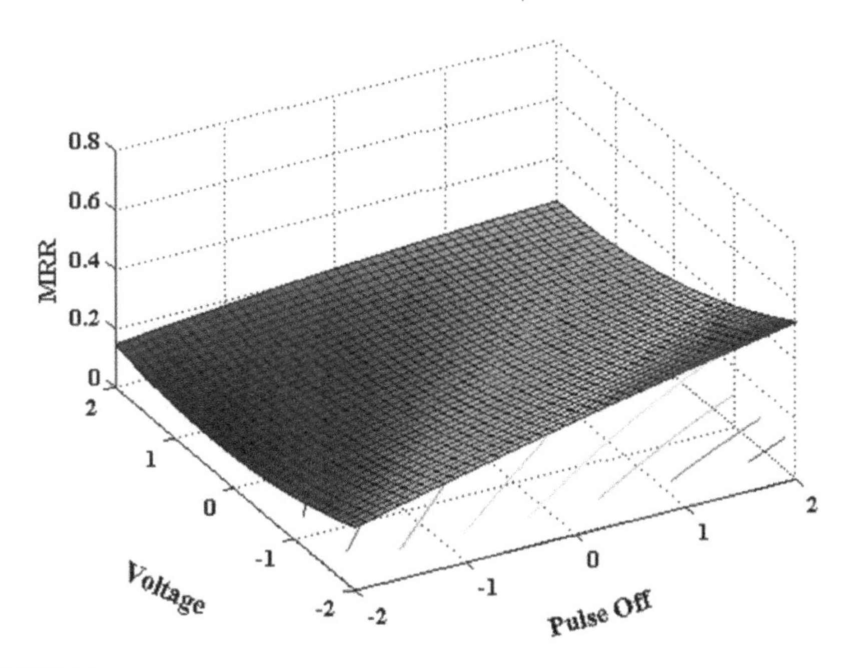

FIGURE 3.9
MRR vs. T_{off} and V.

TABLE 3.5

Experimental Test Result with Mid-Value Predicted Result of the Developed Model

				MRR (g/min)		
T_{on} (μs)	T_{off} (μs)	I_p (A)	V (V)	Exp	Predicted	Error (%)
300(0)	1,700(0)	12(0)	60(0)	0.22752	0.222579	2.17

4

Multi-Objective Optimization Using Weighted Principal Component Analysis

4.1 Introduction

Problems in statistical study involving multi-variables and multi-objective functions are difficult to solve. The multi-responses are sometimes interrelated to each other in addition to their relations with the different variables. Under such circumstances, the problem is made simpler by representing the number of correlated responses with a single performance index. The correlated responses are thus eliminated and are replaced by quality performance indices, which can be either equal or less than the number of their corresponding correlated responses. These quality indices are termed as principal components. The principal components are given their individual accountability proportion (AP) values, and the components having the maximum value are considered. A number of components are sometimes in a considerable range of their AP values (composite components). Under such circumstances, calculations have to be done for solving the composite components.

In this chapter, experiments are performed to find the machining characteristics of electro-discharge machining (EDM) considering material removal rate (MRR) and five other surface roughness values as the responses and voltage, pulse on-time, pulse off-time and current as variables on EN 19 tool steel workpiece with L27 Design of Experiments.

4.2 Experimentation, Result and Discussion

Experiments are performed for obtaining the responses in terms of MRR and other five surface roughness values (Table 4.1).

The quality loss factor is then computed to form a signal-to-noise (S/N) ratio. The S/N ratio for each response-variable relation is calculated with all

TABLE 4.1
Experimental Results

Exp. No.	MRR (g/min)	Ra (µm)	Rku	Rsk	Rsm (mm)	Rq (µm)
1	0.142	9.514	3.580	0.495	0.208	11.540
2	0.198	11.568	3.962	0.378	0.220	14.100
3	0.212	11.980	2.862	0.310	0.243	14.380
4	0.154	9.068	3.340	0.408	0.193	11.042
5	0.208	10.632	3.112	0.257	0.224	12.960
6	0.381	12.040	3.108	0.308	0.246	14.460
7	0.154	9.238	3.002	0.378	0.204	11.260
8	0.389	11.018	3.426	0.457	0.238	13.440
9	0.407	11.310	3.586	0.393	0.223	13.500
10	0.104	9.618	3.494	0.336	0.216	11.720
11	0.163	12.220	3.036	0.263	0.242	14.580
12	0.293	11.968	3.810	0.322	0.238	14.488
13	0.123	10.772	3.240	0.302	0.178	12.940
14	0.318	11.960	3.282	0.108	0.234	14.480
15	0.327	11.710	2.958	0.213	0.218	14.060
16	0.226	11.140	3.084	0.281	0.222	13.520
17	0.310	10.432	3.894	0.303	0.254	12.720
18	0.342	12.300	3.050	0.155	0.254	14.780
19	0.089	10.386	3.846	0.199	0.219	12.520
20	0.240	11.380	3.348	0.110	0.242	13.520
21	0.265	12.540	3.062	0.085	0.236	15.080
22	0.178	10.748	3.406	0.205	0.237	13.070
23	0.259	12.360	3.070	0.198	0.239	14.920
24	0.291	12.680	3.660	0.202	0.260	15.320
25	0.175	9.272	3.346	0.208	0.206	11.200
26	0.253	11.740	3.212	0.141	0.237	14.020
27	0.517	13.140	3.116	0.137	0.256	15.860

the 27 experiments conducted (Table 4.2). It is calculated to be in the span of 0 to 1; the higher the better criteria for MRR, the lower the better criteria for the surface roughness values. Table 4.3 gives the values of scaled (S/N) ratio, and the Pearson's coefficient correlation is given in Table 4.4 for all responses. The non-zero coefficient correlations indicate responses that are correlated to each other. These coefficient correlations are then eliminated by applying principal component analysis using statistical software STATISTICA. The results, for example, eigenvector, eigenvalues, cumulative AP and AP, are then presented (Table 4.5).

The correlated components are then converted to uncorrelated principal components, and equations are derived out of the first, second, third, fourth, fifth and sixth principal components as follows:

TABLE 4.2
S/N Ratio

	S/N Ratio					
Exp. No.	**MRR**	**Ra**	**Rku**	**Rsk**	**Rsm**	**Rq**
1	0.142	–19.567	–11.078	6.104	13.605	–21.244
2	0.198	–21.265	–11.958	8.459	13.128	–22.984
3	0.212	–21.569	–9.133	10.167	12.266	–23.155
4	0.154	–19.150	–10.475	7.791	14.271	–20.861
5	0.208	–20.532	–9.861	11.800	12.987	–22.252
6	0.381	–21.613	–9.850	10.240	12.153	–23.203
7	0.154	–19.312	–9.548	8.446	13.773	–21.031
8	0.389	–20.842	–10.696	6.802	12.468	–22.568
9	0.407	–21.069	–11.092	8.103	13.003	–22.607
10	0.104	–19.662	–10.867	9.477	13.287	–21.379
11	0.163	–21.741	–9.646	11.601	12.302	–23.275
12	0.293	–21.560	–11.619	9.841	12.454	–23.220
13	0.123	–20.646	–10.211	10.406	14.953	–22.239
14	0.318	–21.555	–10.323	19.353	12.601	–23.215
15	0.327	–21.371	–9.420	13.424	13.207	–22.960
16	0.226	–20.938	–9.782	11.041	13.065	–22.620
17	0.310	–20.367	–11.808	10.382	11.890	–22.090
18	0.342	–21.798	–9.686	16.217	11.896	–23.394
19	0.089	–20.329	–11.700	14.041	13.167	–21.952
20	0.240	–21.123	–10.496	19.185	12.309	–22.620
21	0.265	–21.966	–9.720	21.432	12.512	–23.568
22	0.178	–20.627	–10.645	13.783	12.483	–22.326
23	0.259	–21.840	–9.743	14.086	12.425	–23.475
24	0.291	–22.062	–11.270	13.893	11.694	–23.705
25	0.175	–19.344	–10.491	13.656	13.723	–20.984
26	0.253	–21.393	–10.136	17.005	12.498	–22.935
27	0.517	–22.372	–9.872	17.285	11.822	–24.006

$$Z_1^i = -0.429 * Y_{i1} + 0.513 * Y_{i2} + 0.51229 * Y_{i3} - 0.243 * Y_{i4} - 0.149 * Y_{i5} + 0.455 * Y_{i6} \quad (4.1)$$

$$Z_2^i = 0.163 * Y_{i1} - 0.018 * Y_{i2} - 0.047 * Y_{i3} - 0.380 * Y_{i4} - 0.870 * Y_{i5} - 0.260 * Y_{i6} \quad (4.2)$$

$$Z_3^i = -0.256 * Y_{i1} + 0.078 * Y_{i2} + 0.101 * Y_{i3} + 0.863 * Y_{i4} - 0.398 * Y_{i5} - 0.112 * Y_{i6} \quad (4.3)$$

$$Z_4^i = -0.735 * Y_{i1} - 0.444 * Y_{i2} - 0.440 * Y_{i3} - 0.146 * Y_{i4} - 0.097 * Y_{i5} + 0.190 * Y_{i6} \quad (4.4)$$

TABLE 4.3
Scaled S/N Ratio

	Scaled S/N Ratio					
Exp.No.	**MRR**	**Ra**	**Rku**	**Rsk**	**Rsm**	**Rq**
1	0.871	0.878	0.587	0.312	0.266	0.000
2	0.344	0.325	0.440	0.000	0.455	0.069
3	0.249	0.271	0.176	1.000	0.494	0.119
4	1.000	1.000	0.791	0.525	0.312	0.050
5	0.571	0.558	0.397	0.743	0.484	0.167
6	0.236	0.255	0.141	0.746	0.827	0.122
7	0.950	0.946	0.638	0.853	0.311	0.069
8	0.475	0.457	0.238	0.447	0.839	0.021
9	0.404	0.445	0.402	0.307	0.864	0.059
10	0.841	0.835	0.489	0.387	0.088	0.099
11	0.196	0.232	0.187	0.819	0.343	0.162
12	0.252	0.250	0.233	0.120	0.678	0.110
13	0.536	0.562	1.000	0.619	0.184	0.126
14	0.254	0.251	0.278	0.579	0.724	0.389
15	0.311	0.333	0.464	0.899	0.740	0.215
16	0.445	0.441	0.421	0.770	0.530	0.145
17	0.622	0.609	0.060	0.053	0.710	0.126
18	0.178	0.195	0.062	0.804	0.765	0.297
19	0.634	0.653	0.452	0.091	0.000	0.233
20	0.388	0.441	0.189	0.518	0.564	1.000
21	0.126	0.139	0.251	0.792	0.622	0.451
22	0.542	0.534	0.242	0.465	0.394	0.226
23	0.165	0.169	0.224	0.784	0.608	0.235
24	0.096	0.096	0.000	0.244	0.673	0.229
25	0.940	0.961	0.623	0.520	0.386	0.222
26	0.304	0.341	0.247	0.645	0.594	0.320
27	0.000	0.000	0.039	0.739	1.000	0.329

$$Z_5^i = 0.426 * Y_{i1} - 0.188 * Y_{i2} - 0.171 * Y_{i3} + 0.166 * Y_{i4} - 0.225 * Y_{i5} + 0.821 * Y_{i6} \tag{4.5}$$

$$Z_6^i = 0.003 * Y_{i1} - 0.705 * Y_{i2} + 0.708 * Y_{i3} - 0.023 * Y_{i4} - 0.008 * Y_{i5} - 0.014 * Y_{i6} \tag{4.6}$$

where Y_{i1}, Y_{i2}, Y_{i3}, Y_{i4}, Y_{i5} and Y_{i6} are the scaled S/N ratio values of material removal rate and surface roughness, respectively, for the *i*th trial.

The multi-response performance index (MPI) for the *i*th trial in the form of weighted sum of the principal responses is calculated as follows:

TABLE 4.4
Pearson's Correlation Coefficient amongst Responses

Sl. No.	Responses	Pearson's Correlation Coefficient	Remarks
1	Ra & Rq	0.998	Both correlated
2	Ra & Rsk	–0.342	Both correlated
3	Ra & Rku	–0.241	Both correlated
4	Ra & Rsm	0.713	Both correlated
5	Ra & MRR	–0.653	Both correlated
6	Rq & Rsk	–0.313	Both correlated
7	Rq & Rku	–0.224	Both correlated
8	Rq & Rsm	0.722	Both correlated
9	Rq & MRR	–0.661	Both correlated
10	Rsk & Rku	0.163	Both correlated
11	Rsk & Rsm	–0.333	Both correlated
12	Rsk & MRR	0.173	Both correlated
13	Rku & Rsm	–0.025	Both correlated
14	Rku & MRR	0.165	Both correlated
15	Rsm & MRR	–0.646	Both correlated

TABLE 4.5
Computed Various Principal Component Analysis (PAC) Values

	MRR	Ra	Rku	Rsk	Rsm	Rq
Eigenvector	–0.430	0.513	–0.150	–0.244	0.456	0.513
	0.164	–0.019	–0.871	–0.380	–0.261	–0.047
	–0.257	0.078	–0.399	0.864	–0.113	0.101
	–0.736	–0.444	–0.098	–0.147	0.191	–0.440
	0.426	–0.189	–0.226	0.167	0.822	–0.171
	0.004	–0.705	–0.008	–0.023	–0.014	0.708
Eigenvalues	3.410	1.015	0.290	0.434	0.002	0.849
AP	0.568	0.169	0.048	0.072	0.000	0.142
CAP (cumulative accountability proportion)	0.568	0.738	1.000	0.951	1.000	0.879

$$MPI^i = 0.5684 * Z_1^i + 0.1692 * Z_2^i + 0.1415 * Z_3^i + 0.0723 * Z_4^i + 0.0483 * Z_5^i + 0.003 * Z_6^i \quad (4.7)$$

The MPI factors for 27 experimentations are enlisted in Table 4.6, and the level averages for each controlling factors are summarized in Table 4.7. The average of all MPI values corresponding to the first level of the control factor A is the level average on the MPI for factor A at level 1. Larger value of MPI has higher significance likewise in Figure 4.1 it was seen that the plot

TABLE 4.6
Evaluated MPI Values

Exp.	A	B	C	D	MPI
1	1	1	1	1	0.428
2	1	1	1	1	0.148
3	1	1	1	1	–0.279
4	1	2	2	2	0.461
5	1	2	2	2	0.011
6	1	2	2	2	–0.313
7	1	3	3	3	0.294
8	1	3	3	3	–0.078
9	1	3	3	3	–0.025
10	2	1	2	3	0.414
11	2	1	2	3	–0.205
12	2	1	2	3	–0.051
13	2	2	3	1	0.283
14	2	2	3	1	–0.216
15	2	2	3	1	–0.222
16	2	3	1	2	–0.065
17	2	3	1	2	0.103
18	2	3	1	2	–0.378
19	3	1	3	2	0.399
20	3	1	3	2	–0.142
21	3	1	3	2	–0.326
22	3	2	1	3	0.065
23	3	2	1	3	–0.291
24	3	2	1	3	–0.237
25	3	3	2	1	0.359
26	3	3	2	1	–0.166
27	3	3	2	1	–0.529

TABLE 4.7
Level Average on MPI

Factors	Level 1	Level 2	Level 3
A	0.0719	–0.0385	–0.0964
B	0.0416	–0.0508	–0.0538
C	0.2919	–0.0927	–0.2623
D	–0.0216	0.068	–0.0309

for the process parameters (pulse on-time and current) has the highest inclination, and these parameters are considered as the most significant parameters. The other two parameters (pulse off-time and voltage) also contribute a small amount of MPI but are less significant. Thus, the optimal condition

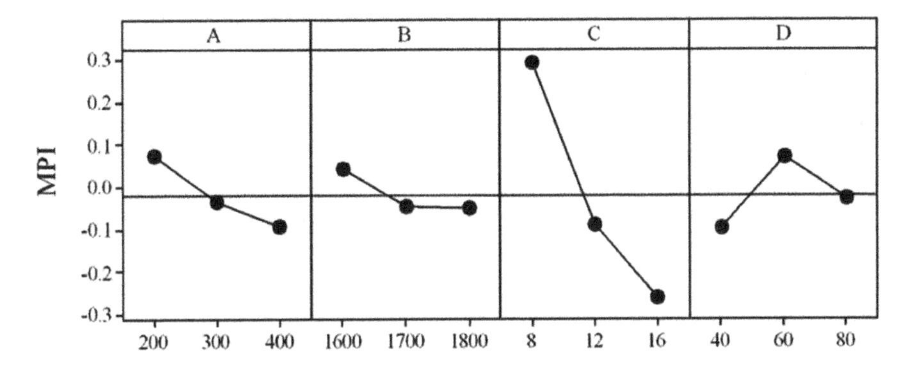

FIGURE 4.1
Mean effects plot for MPI.

for the process parameters (A–D) can be set as A1B1C1D2, which denotes (lowest levels of pulse on-time, pulse off-time and current and midlevel of voltage).

4.3 Analysis of Variance

In Table 4.8, analysis of variance (ANOVA) test is performed, which determines the effect of design parameters on the MPI value. It was found that the current (C) has the highest contribution in effectiveness (70%) in determining the multiple responses and is considered the most significant parameter.

TABLE 4.8
ANOVA Test Results

Factors	DOF	SS	MS	F-Ratio	Contribution (%)
A	2	0.132	0.066	1.98	6.31
B	2	0.053	0.027	0.8	2.54
C	2	1.452	0.726	21.88[a]	69.68
D	2	0.129	0.064	1.94	6.17
A*B	4	0.018	0.004	0.13	0.84
A*C	4	0.047	0.012	0.35	2.25
B*C	4	0.055	0.014	0.41	2.63
Error	6	0.199	0.033		9.55
Total	26	2.083			100

[a] Significant at 95% confidence level ($F_{0.05,2,6} = 5.14$).

4.4 Confirmation Test

Table 4.9 shows the comparison between the initial (S/N) ratio and the actual (S/N) ratio to detect the accuracy of the analysis considering the optimal parameters.

The increase of the optimal (S/N) ratio from the initial (S/N) ratio is found to be 1.655, which denotes an improvement of 4% in responses. Thus, the experimental result validates the design and analysis for optimal solution of machining parameters.

TABLE 4.9

Results of Confirmatory or Validation Test

Responses	Mid-Level Combination (A2B2C2D2)	Optimal Combination (A1B1C1D2)
Ra	10.496	9.514
Rq	13.260	11.540
Rsk	0.410	0.495
Rku	3.628	3.280
Rsm	213.000	195.000
MRR	0.248	0.252
S/N ratio	–44.451	–42.796

Note: Enhancement in S/N ratio = 1.655 dB, i.e., 4%.

5

Single-Objective Optimization Using Taguchi Technique (Maximization)

5.1 Introduction

Lathe machining is an old technique used for cutting cylindrical workpieces by rotating it against a cutting tool held between two adjacent trees. In 1797, an Englishmen named Henry Maudslay developed the first screw-cutting lathe machine, which is the origin of the modern high-speed, heavy-duty machining. The machine involves a cylindrical work part fixed to a spindle by a chuck and a tool header that holds the tool rigidly and is contacted against the circumference of the work part to cut it to a required shape and size. Computer numerical control (CNC) machining is another machining technology that consists of a machining centre run by a computerized system. This study considers the material removal rate (MRR) as a machining response that is to be evaluated for high production rate and reduced cost. The cutting conditions (cutting speed, feed rate and depth of cut) were taken as variables that need to be optimized. The objective is to maximize the MRR with the most optimal solution of the machining parameters (cutting conditions) using Taguchi method.

5.2 Experimentation, Result and Discussion

The CNC machine tool (EMCO Concept Turn 105) is used for the purpose of experiment with a carbide tool and mild steel workpiece (Society of Automotive Engineers [SAE] 1020) combination, and the cutting parameters are cutting speed (A), feed rate (B) and depth of cut (C). Five levels of these input parameters are taken, equispaced within the limits of their operating range. Taguchi method has been applied in experimentation using L25 design of experiments (DOE), consisting of 25 experiments with different levels and combination of input parameters. The cutting parameters in five different levels of their operating range are shown in Table 5.1.

TABLE 5.1

Process Parameters and Their Levels

Parameters	Code	Levels				
		1	2	3	4	5
Speed (m/s)	A	6	6.2	6.4	6.6	6.8
Feed (mm/rev)	B	1.5	2	2.5	3	3.5
Depth of cut (mm)	C	1	1.5	2	2.5	3

The Taguchi methodology enables extracting the optimal cutting conditions based on L25 orthogonal array, which gives the most economical MRR. The better criteria for the MRR are applied considering 25 experiments and the S/N ratio for each experimental process parameter is evaluated (Table 5.2). The average S/N value for each level of process parameters is

TABLE 5.2

Tabulation of Parameter Levels and Outputs

Exp. No.	Speed (m/s) (A)	Feed (mm/rev) (B)	Depth of Cut (mm) (C)	MRR (mm^3/min)	S/N Ratio (dB)
1	1	1	1	1,514.51	79.73
2	1	2	2	2,173.64	83.80
3	1	3	3	3,873.10	90.33
4	1	4	4	4,647.66	92.39
5	1	5	5	5,690.31	94.67
6	2	1	2	1,804.00	81.70
7	2	2	3	2,883.15	86.99
8	2	3	4	4,795.62	92.74
9	2	4	5	5,516.13	94.32
10	2	5	1	2,250.30	84.20
11	3	1	3	2,846.52	86.85
12	3	2	4	3,960.68	90.58
13	3	3	5	5,154.60	93.55
14	3	4	1	2,734.86	86.40
15	3	5	2	3,767.97	90.02
16	4	1	4	3,188.42	88.13
17	4	2	5	4,588.90	92.24
18	4	3	1	1,924.23	82.43
19	4	4	2	3,839.95	90.23
20	4	5	3	5,073.21	93.38
21	5	1	5	3,676.70	89.74
22	5	2	1	2,112.98	83.48
23	5	3	2	2,860.25	86.90
24	5	4	3	4,482.01	91.98
25	5	5	4	5,839.68	94.96

TABLE 5.3

Responses of Mean S/N Ratio

Level	A	B	C
1	69.86	67.53	65.96
2	69.71	69.27	68.56
3	70.89	70.67	71.23
4	70.74	72.15	72.70
5	70.84	72.45	73.61
Delta	1.18	4.92	7.65
Rank	3	2	1

Note: The total mean S/N ratio = 70.41 dB.

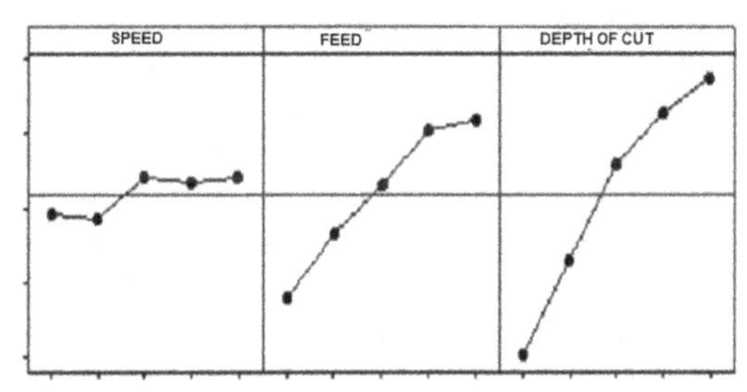

FIGURE 5.1
Mean effect plot for variation of S/N ratio.

then calculated and presented in Table 5.3. Calculations are done using a Minitab software. Ranks based on delta method are assigned by considering the difference in the average S/N values between the highest and the lowest for each factor. The effect plots of input parameters are shown in Figure 5.1.

When we observe the mean effect plot between the S/N ratio and the process parameter values, we find the speed plot exhibiting a horizontal plot that indicates a factor with a negligible significance. The feed rate and the depth of cut plots, on the other hand, exhibit a higher inclination, indicating a higher significance. The optimal parameters are selected by considering the values of the parameter in each level having the highest S/N value. Thus, the maximum MRR can be represented as A3B5C5.

5.3 Analysis of Variance

Analysis of variance (ANOVA) is a statistical technique that determines the conclusion as to which combination of parameters is to be chosen for a better

TABLE 5.4
ANOVA Output

Source of Variation	Degree of Freedom	Sum of Squares	Mean Square	F-Ratio	Contribution (%)
Speed	4	222,328	55,582	0.61	0.89
Feed	4	7,449,375	1,880,602	20.41	29.97
Depth of cut	4	16,047,106	4,051,107	43.98	64.56
Error	12	1,138,511	94,875		4.58
Total	24	24,857,319			100

response (maximum MRR in this case). It analyses the data and forms an F factor that indicates the relationship of regression mean square to mean square error. It is also known as the variance ratio, as it represents response variation caused because of change in values of factors or just the change in error range. F factor has a definite role to play in determining the significant parameters by comparing the variances in error terms having the significance level (α) with the already-tabulated standard value. The factors having the higher calculated value in variance than the tabulated value will be considered the significant one. F factor represents a criterion where a higher value indicates a greater significance in the factor. The percentage contribution of each parameters effecting the MRR is depicted in the ANOVA table (Table 5.4), which complies with the main effect plot previously discussed and shows the significance of the cutting parameters based on the S/N value inclination, i.e., B and C having the most significance and A having the least significance within the specific range of values.

5.4 Confirmation Test

Lastly, one more confirmation with the experimentation being performed taking the mid-term values of the process parameters is represented as A3B3C3 and is compared with the optimal parameter values (A3B5C5). Table 5.5 shows the data obtained from the confirmation test, where the

TABLE 5.5
Results of Confirmatory or Validation Test

		Optimized Values	
	Mid-Term Values	Theoretical	Experimental
Combination	(A3)(B3)(C3)	(A3)(B5)(C5)	(A3)(B5)(C5)
MRR	2,931.20		4,646.62
S/N ratio (dB)	71.15	75.65	76.33

Note: Improvement of S/N ratio = 4.37 dB (6.33%).

actual S/N ratio value is compared with the theoretical value, and the variation is found to be 4.37dB, which is 6.33% improvement in responses. This implies that the predicted value is in compliance with the experimental results, which signifies the reproducibility of the experiment.

6

Single-Objective Optimization Using Taguchi Technique (Minimization)

6.1 Introduction

Electrochemical machining (ECM) is a non-conventional machining technique widely used in the modern manufacturing sector. For industrial and commercial reasons, ECM is required to have minimal undesirable effects such as overcut (OC) and surface roughness with the most achievable optimum parameter. OC is the main element of error in ECM, and the parameters involved need to be recognized.

6.2 Experimentation

The experiment is carried out for EN19 tool steel, which involves Electrolyte Concentration (%), Feed rate (mm/min), Voltage (V) and Inter-electrode gap (mm) as the process parameters and OC as the response. OC is taken as the variation in tool and cut dimensions (Figure 6.1). The optimum levels in parameters are determined for minimum OC, and Taguchi method has been

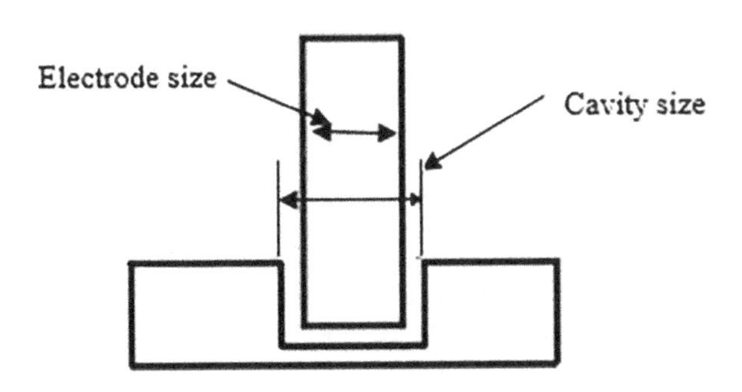

FIGURE 6.1
Systematic diagram of an overcut.

TABLE 6.1
Input Table

Input Parameters	Symbols	Levels 1	2	3
Electrolyte concentration	A	20	25	30
Applied voltage	B	10	12	14
Rate of feed	C	0.2	0.4	0.6
Gap	D	0.15	0.2	0.25

applied in the evaluation. Three levels of each parameter are taken, i.e., three equidistant values as shown in Table 6.1.

$$\text{Overcut} = \frac{(\text{size of cavity produce on workpiece} - \text{size of electrode})}{2} \quad (6.1)$$

6.3 Result and Discussion

Experiment was performed using Taguchi method of optimization mentioned earlier, with 27 experiments (L27) conducted for a different set of combination of input parameters compared at different levels. From each set of combination of parameters, the corresponding signal-to-noise (S/N) value is computed using a Minitab 16 software (Table 6.2). The effect of the input parameters can be felt from the response table evaluated from the mean S/N values in each level (Table 6.3). Delta values are evaluated considering the difference in the maximum and minimum values of the parameter and ranks are given accordingly. The parameter possessing the highest value of delta will yield the first rank.

6.4 Plots

For all S/N ratios and mean values, plot are generated for each level of the input parameters (Figure 6.2) using a Minitab software. The main effect plots indicated that with an increase in voltage, feed rate and concentration OC decrease. The main effect plot also suggests a decrease in S/N ratio with an increase in voltage and concentration. The significance of the parameters can easily be found out by observing the inclination of the parameters.

TABLE 6.2
Design of Experiment Table and Result

Trial	A	B	C	D	OC	S/N
1	1	1	1	1	1.52	–3.673
2	1	2	2	3	1.27	–2.097
3	1	1	2	2	1.41	–3.014
4	1	2	1	2	1.45	–3.259
5	1	1	3	3	1.38	–2.826
6	1	2	3	1	1.69	–4.604
7	1	3	2	1	1.62	–4.232
8	1	3	3	2	1.86	–5.444
9	1	3	1	3	2.21	–6.957
10	2	1	3	1	1.32	–2.435
11	2	1	2	3	1.51	–3.616
12	2	1	1	2	1.48	–3.439
13	2	2	2	1	2.45	–7.861
14	2	2	3	2	2.58	–8.314
15	2	2	1	3	2.32	–7.383
16	2	3	2	2	1.82	–5.253
17	2	3	1	1	1.7	–4.655
18	2	3	3	3	2.34	–7.458
19	3	1	3	2	1.64	–4.340
20	3	1	2	1	1.6	–4.123
21	3	1	1	3	1.46	–3.320
22	3	2	3	3	1.82	–5.253
23	3	3	3	1	2.78	–8.970
24	3	2	2	2	1.66	–4.446
25	3	2	1	1	1.33	–2.502
26	3	3	2	3	2.54	–8.178
27	3	3	1	2	2.2	–6.916

TABLE 6.3
Response Table

Levels/Parameters	A	B	C	D
First	–4.766	–3.726	–2.779	–0.332
Second	–6.655	–5.533	–2.827	–0.342
Third	–6.343	–7.027	–3.277	–0.363
Delta	1.889	3.301	0.497	0.031
Rank	2	1	3	4

The higher the inclination, the higher the significance of the parameters. Thus, the optimal parameter values can be represented as A1B1C1D1 for ECM in the assumed level of operation.

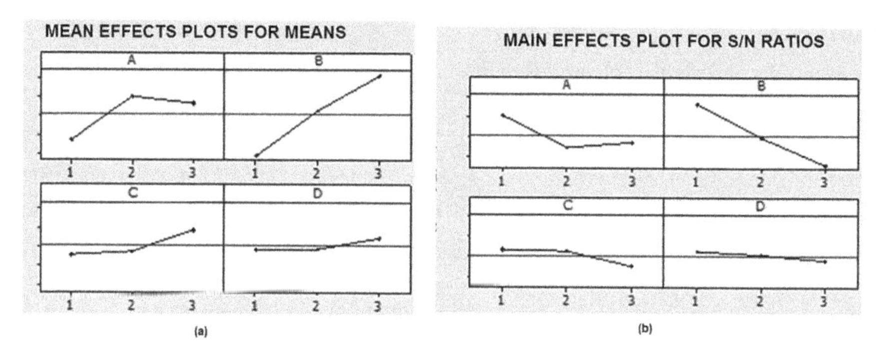

FIGURE 6.2
Main effect plot: (a) means; (b) S/N ratios.

6.5 Analysis of Variance

Analysis of variance (ANOVA; Table 6.4) is computed for analysis of the input parameters using a Minitab software. The table is based on an *F* value that is taken as the ratio of mean square of variance to mean square of error, and the whole result is calculated at 95% confidence level.

Standard *F* values having higher values than calculated indicate a higher significance of that parameter. From the ANOVA table, *F* values computed indicate a higher significance of voltage on the OC response, which is measured at a confidence level of 95%. Voltage is being followed by concentration level, the second significant factor. From the % contribution, it is seen that voltage has a contribution of 48%, while concentration has 16.15%

TABLE 6.4
ANOVA Table

Input and Interactions	Degrees of Freedom	Seq.Sum of Squares	Adjusted Sum of Squares	Adjusted Mean Square	*F*	*p*	Contribution (%)
A	2	0.634	0.634	0.317	13.996	0.009	16.147
B	2	1.884	1.884	0.942	41.607	0.000	48.003
C	2	0.204	0.204	0.102	4.520	0.091	5.214
D	2	0.048	0.048	0.024	1.070	0.500	1.234
AB	4	1.825	1.825	0.456	20.148	0.002	23.245
BC	4	0.179	0.179	0.045	1.978	0.295	2.282
AC	4	0.304	0.304	0.076	3.358	0.131	3.874
Error	6	0.156	0.156	0.026			
Total	26	5.235					

Note: S = 0.160, R-sq (actual) = 95.05% and R-sq (adjusted) = 86.04%.

contribution for OC response. The percentage contribution of the interaction between voltage and concentration on material removal rate has been taken as 34.85%.

6.6 Development of Regression Model for OC

Regression equation has been formed using the Minitab software to find the relation between the input parameters and the responses. It is given as follows:

$$\begin{aligned}\text{Overcut} = {} & (-0.374)*\text{A}+(-0.309)*\text{B}+(0.070)*\text{C}+(0.316)*\text{D} \\ & +((0.166)*(\text{A}*\text{B}))+((0.086)*(\text{B}*\text{C}))+((-0.182)*(\text{C}*\text{D})) \\ & +((0.109)*(\text{A}*\text{C}))+((-0.015)*(\text{A}*\text{D}))+((0.063)*(\text{B}*\text{D}))+(1.503)\end{aligned} \tag{6.2}$$

6.7 Confirmatory Experiment

The predicted and experimental S/N values are compared to check the improvement in results in Table 6.5. The predicted S/N values are calculated using Minitab, and the experimental values are obtained as follows:

$$\tilde{\gamma} = \gamma_m + \sum_{i=1}^{o}\left(\tilde{\gamma}_i - \gamma_m\right) \tag{6.3}$$

where $\tilde{\gamma}$ is the best optimized value of individual input parameter, γ_m is the total sum of the mean S/N ratio, $\tilde{\gamma}_i$ is the mean S/N ratio and o is the number

TABLE 6.5
Validation Table

Outputs	Mid-Value	Optimal Set	
		Value Predicted	**Value Obtained**
		A2B2C2D2	A1B1C1D1
S/N ratio	–9.564	–4.890	–4.546
OC	3.009		1.900

Note: Improvement in S/N ratio = (–4.546)–(–9.564) = 5.017 dB.

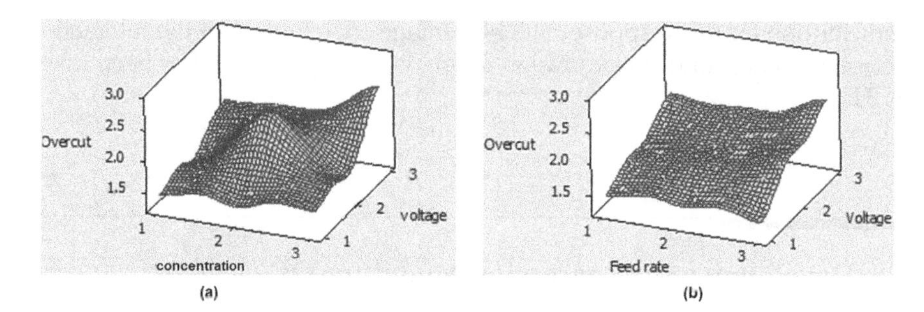

FIGURE 6.3
Three-dimensional plot: (a) overcut with voltage and concentration; (b) overcut with voltage and feed rate.

of input parameters. The small value of the difference between the predicted and experimental values indicates a good conformation of the optimized parameter response value with the experimental values. The increment in the S/N ratio is evaluated as 5.017 dB, as indicated in Table 6.5.

6.8 Three-Dimensional Plots

Surface plots shown in Figure 6.3a define the OC response as a function of voltage and concentration parameters. The OC response increases with an increase in voltage and concentration. The second plot (Figure 6.3b) exhibits an increase in OC with a decrease in inter-electrode gap as well as feed rate.

6.9 Conclusion

In this chapter, input parameters (electrolyte concentration, applied voltage, rate of feed, and gap) have been optimized, and the response results are compared using Taguchi technique. S/N ratios, regression models and ANOVA tables are constructed for the study of behaviour in response with the parameters. It has been found that the voltage and concentration have the highest and second highest influence on the OC and the interactive influence on the voltage and concentration, and also an effect on the OC. Thus, the optimized values of the parameters are given as electrolyte concentration (20%), applied voltage (10V), rate of Feed (0.2 mm/min) and gap (0.15 mm).

7

Multi-Objective Optimization Using Grey-Taguchi Technique

7.1 Introduction

For critical weight applications in automotive and space industries, the need of materials having higher strength-to-weight ratio, maintaining the toughness and wear resistance at the same time, is preferred. In recent times, the use of polymeric composites has become quite evident due to the possession of all the aforementioned desired properties. Though these materials are suitable for use, in some applications where production cost is of main concern, cheaper filler materials are employed with the polymeric composite to get the desired quality. As these materials are used in applications where sliding motion can be prevalent, wear resistance is considered as the most significant response in determining its quality, including friction coefficient. The sample for the present experiment includes a compression moulded mixture of resin (Araldite AW106) and hardener (Araldite HV953U) in the ratio of 10:8 by weight incorporated with kaolin fillers, which is a hydrated aluminium silicate mineral with an average particle size of 564 nm. These samples were then cut in dimensions of $20 * 20 * 8\,mm^3$ for tribological testing.

7.2 Experimentation, Result and Discussion

L27 design of experiments (DOE) has been applied, with particulate filler (weight percent), applied load (Newton) and sliding speed revolution per minute (RPM) being the input factors each having three levels, which is shown in Table 7.1.

Tribological tests are performed involving a block-on-roller arrangement with applied normal load over the specimen for a constant period of time

TABLE 7.1
Process Parameters and their Levels

Process Parameters	Code	Unit	Levels		
			1	2	3
Particulate filler	A	%	1.5	4.5	7.5
Normal load	B	N	20	30	40
Sliding speed	C	rev/min	85	95	105

(300 s) under dry conditions (ambient temperature is about 28°C with a relative humidity at 85%). Roller made of EN8 steel (55HRC [Rockwell hardness measured on the C scale) having 50 mm diameter and 50 mm thickness is used. After cleaning the mating surfaces, the specimen is pushed hard against the roller with a specified normal load. Load, with the aid of a lever, is applied over the specimen by placing dead weights, and load-measuring sensors provided in the machine measure the normal load. The coefficient of friction is measured and displayed on the computer, and the wear rate is calculated by measuring the weight loss of the composite after each experimental operation. Table 7.2 provides the experimental results.

7.3 Grey Relational Analysis

A normalization technique is used for obtaining a single grade of normalized responses for optimization of the multi-variable system. Since the coefficient of friction and wear rate need to be minimized, a normalization technique of lower the better criteria is selected. A grey relational coefficient is extracted out of the normalized responses, which denotes the weighted share of the individual responses. The relationship between the actual and ideal values can also be developed from the Grey relational coefficients and grades (Table 7.3).

7.4 Grey Relational Grade Analysis

Grey relational grades are obtained by analysing the grey relational coefficients and evaluating the sum of their weightages. It also describes the cumulative effect of each response. The higher the value of the relational grade, the better the relation between the responses in their given and ideal values.

TABLE 7.2

Experimental Results for Response Parameter

Exp. No.	Particulate Filler (A)	Normal Load (B)	Sliding Speed (C)	Friction Coefficient	Sp. Wear Rate (mm^3/Nm)
1	1	1	1	0.294	0.0002
2	1	1	2	0.160	0.0010
3	1	1	3	0.189	0.0005
4	1	2	1	0.122	0.0005
5	1	2	2	0.162	0.0005
6	1	2	3	0.157	0.0006
7	1	3	1	0.135	0.0004
8	1	3	2	0.165	0.0005
9	1	3	3	0.155	0.0006
10	2	1	1	0.344	0.0025
11	2	1	2	0.370	0.0043
12	2	1	3	0.307	0.0033
13	2	2	1	0.277	0.0037
14	2	2	2	0.265	0.0023
15	2	2	3	0.273	0.0020
16	2	3	1	0.261	0.0013
17	2	3	2	0.250	0.0029
18	2	3	3	0.215	0.0015
19	3	1	1	0.399	0.0020
20	3	1	2	0.471	0.0024
21	3	1	3	0.356	0.0028
22	3	2	1	0.349	0.0020
23	3	2	2	0.276	0.0032
24	3	2	3	0.239	0.0017
25	3	3	1	0.223	0.0017
26	3	3	2	0.238	0.0018
27	3	3	3	0.175	0.0011

7.5 Taguchi Analysis for Grades

As the multi-responses are now converted into a single relational grade, single response optimization technique can be implemented using the Taguchi method. Asignal-to-noise (S/N) value is computed considering the relational grades by implementing the lower the better criteria for both the outputs (Table 7.4). From the response table given in Table 7.5, it is observed that the input factor having the highest significance is the filler content parameter that affects the responses in wear rate and coefficient of friction to the highest degree. Normal loads and rubbing speed parameters have the lesser

TABLE 7.3

Normalized Values of Experiment Data and Grey Relation Coefficient

	Normalized Data		Grey Relation Coefficient		
Exp. No.	Friction Coefficient	Wear Rate (mm^3/Nm)	Friction Coefficient	Wear Rate (mm^3/Nm)	Grade
1	0.507	1.000	0.504	1.000	0.752
2	0.891	0.797	0.821	0.711	0.766
3	0.808	0.936	0.723	0.886	0.804
4	1.000	0.932	1.000	0.881	0.941
5	0.885	0.921	0.814	0.863	0.838
6	0.900	0.913	0.833	0.852	0.843
7	0.963	0.945	0.931	0.900	0.916
8	0.877	0.916	0.802	0.856	0.829
9	0.905	0.897	0.841	0.829	0.835
10	0.364	0.439	0.440	0.471	0.456
11	0.289	0.000	0.413	0.333	0.373
12	0.470	0.237	0.485	0.396	0.441
13	0.556	0.154	0.530	0.372	0.451
14	0.590	0.495	0.550	0.498	0.524
15	0.567	0.562	0.536	0.533	0.534
16	0.602	0.719	0.557	0.641	0.599
17	0.633	0.336	0.577	0.429	0.503
18	0.734	0.688	0.652	0.616	0.634
19	0.206	0.561	0.386	0.532	0.459
20	0.000	0.454	0.333	0.478	0.406
21	0.330	0.357	0.427	0.438	0.432
22	0.350	0.565	0.435	0.535	0.485
23	0.559	0.264	0.531	0.405	0.468
24	0.665	0.634	0.599	0.577	0.588
25	0.711	0.640	0.633	0.581	0.607
26	0.668	0.619	0.601	0.568	0.584
27	0.848	0.788	0.767	0.702	0.735

significance. Main effect plot and interaction plot for the S/N ratio obtained by Taguchi analysis are shown in Figures 7.1 and 7.2, respectively.

7.6 Analysis of Variance for Grades

Analysis of variance (ANOVA) is performed with the objective of validating whether different population of input factors result in a common mean value or not. It analyses the effect of the factors at each level on the mean value of

TABLE 7.4
Grades and Corresponding S/N Ratios

Exp. No.	Grade	Order	S/N Ratio
1	0.752	19	2.478
2	0.766	20	2.313
3	0.804	21	1.890
4	0.941	27	0.533
5	0.838	24	1.533
6	0.843	25	1.488
7	0.916	26	0.767
8	0.829	22	1.629
9	0.835	23	1.568
10	0.456	6	6.827
11	0.373	1	8.562
12	0.441	4	7.118
13	0.451	5	6.925
14	0.524	11	5.620
15	0.534	12	5.442
16	0.599	15	4.457
17	0.503	10	5.966
18	0.634	17	3.959
19	0.459	7	6.756
20	0.406	2	7.835
21	0.432	3	7.283
22	0.485	9	6.290
23	0.468	8	6.597
24	0.588	14	4.614
25	0.607	16	4.333
26	0.584	13	4.669
27	0.735	18	2.677

TABLE 7.5
Response Table

Level	Filler Content	Load	Rubbing Speed
1	1.578	5.673	4.374
2	6.097	4.338	4.969
3	5.673	3.336	4.004
Delta	4.52	2.337	0.965
Rank	1	2	3

their corresponding responses. The percentage contribution of each factor was calculated by considering the sum of squares of the deviations in S/N value of all factors from the mean S/N value (Table 7.6). The effect of the filler content on both the outputs is thus found out to be the most significant.

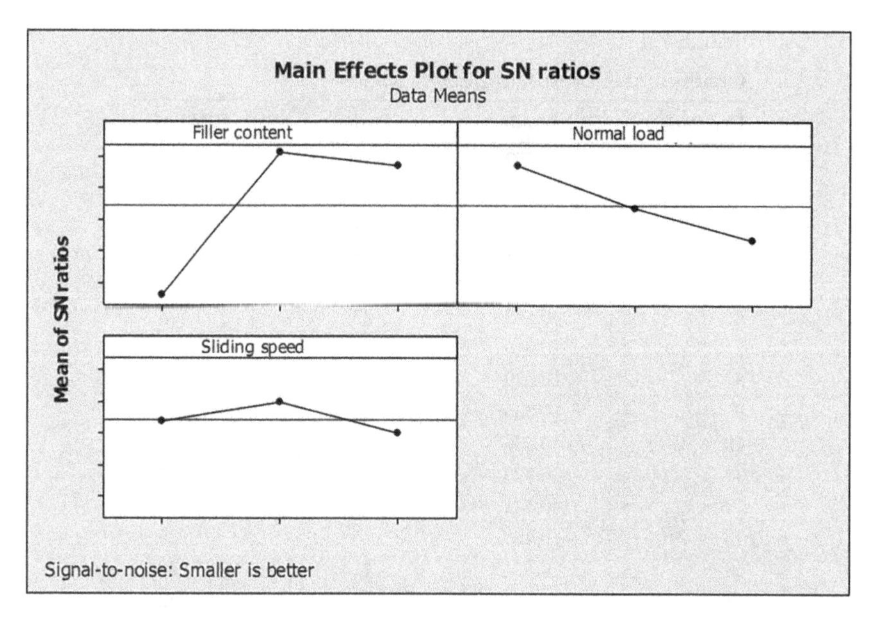

FIGURE 7.1
Main effect plot for the S/N ratio of grey relational grade.

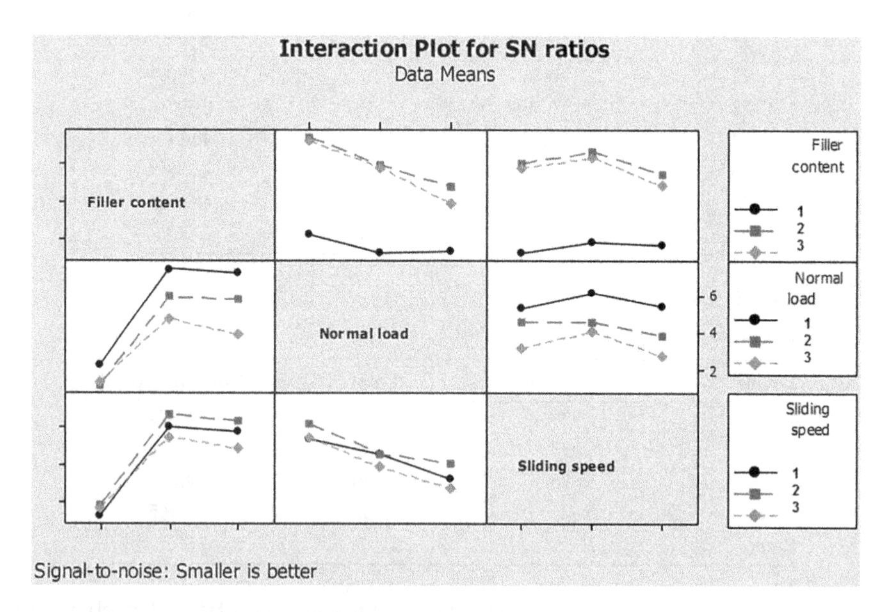

FIGURE 7.2
Interaction plot for the S/N ratio of grey relational grade.

TABLE 7.6

ANOVA for Grades Using Adjusted Sum of Squares for Test

Source of Variation	Degree of Freedom	Sum of Squares	Adjusted Sum of Squares	Adjusted Mean Square	F-Ratio	Contribution (%)
A	2	0.6196	0.6196	0.3098	115.72	0
B	2	0.102329	0.1023	0.0511	19.11	0.001
C	2	0.017773	0.0177	0.0088	3.32	0.089
A*B	4	0.018348	0.0183	0.0045	1.71	0.239
B*C	4	0.003568	0.0035	0.0008	0.33	0.848
A*C	4	0.010312	0.0103	0.0025	0.96	0.477
Error	8	0.021418	0.0214	0.0026		
Total	26	0.793347				

TABLE 7.7

Confirmation Table

		Optimal Parameter	
Parameter	Mean Parameter	Predicted	Experimental
Combination of testing parameters	A2B2C2	A3B3C3	A2B1C2
S/N ratio for coefficient of friction	11.535		8.636
S/N ratio for wear rate	52.883		47.329
S/N ratio for grey relational grade	5.620	2.677	8.562

7.7 Confirmation Test

Confirmation test is being performed to validate the experimental responses with the predicted responses (Table 7.7).

Because there are three levels for each factor, the second level of the filler content is considered for experimentation, which is 4.5%. The difference between the experimental and predicted S/N values was found to be 2.942345 dB, which validates the design and analysis method presented.

8

Multi-Objective Optimization Using Taguchi-Coupled Fuzzy Logic

8.1 Introduction

Multi-response problems are difficult to optimize compared with single-response optimization. In this research work, wear characteristic has been analysed, taking wear resistance and coefficient of friction as the responses for an aluminium composite. Both of these responses are based on the lower the better criteria, and Taguchi method has been used with fuzzy logic for the evaluation of optimization. The following steps have been followed for the evaluation of optimization:

1. Planning of the test in accordance with the Taguchi method.
2. Evaluation of the signal-to-noise (S/N) value, which gives the relation between the evaluated response of the wear characteristic and the preferred response.
3. Loss function for the responses are fuzzified logically and by applying fuzzy rules, and multi-response performance index (MRPI) value is selected for carrying out optimization of the wear characteristic.
4. Analysis of the tests using MRPI and analysis of variance.
5. Evaluation of the optimized parameter.
6. Verification of the result using tests.

8.2 Experimental Design

The Taguchi method of design of experiments (DOE) has been applied in this experiment to evaluate the optimized parameters for better performance of wear characteristics. Taguchi method enables design engineers to carry out systematic evaluation of all design parameters for the responses that best

TABLE 8.1
Data Table for the Input Parameters

Input Parameters	Unit	Levels		
		1	2	3
Volume fraction of reinforcement (A)	%Weight	1.5	5.5	9.5
Load (B)	Newton	100	150	200
Sliding/rubbing speed (C)	Rev/minute	200	300	400

define the high performance of test results. It helps in significantly reducing the number of experimental data required for optimization of performances in response result.

8.3 Experimental Parameters

Before implementing the Taguchi three-level design method, independent input parameters are recognized in a way that they have the highest tendency to influence the responses already mentioned. The parameters taken were the volume fraction of the reinforcement (A), Applied Load (B) and sliding/rubbing speed (C). The parameters along with their symbols and levels are provided in Table 8.1.

8.4 Planning the Experiments

Taguchi's L27 orthogonal array technique has been implemented in this experiment. Degrees of freedom (DOF) of (6×2) has been taken, with the three parameters covering six DOFs and the interacting factors between the parameters covering the remaining DOFs. The L27 orthogonal array is shown in Table 8.2.

8.5 S/N Ratio

An S/N ratio value has been given to each combination of process parameters for the best detection of the process performance that would yield a better criterion to model the relation between the response and the input parameters. It represents the process performance in the presence of various

TABLE 8.2

Orthogonal Array L27 of Taguchi

Test No.	A	B	C
1	1	1	1
2	1	1	2
3	1	1	3
4	1	2	1
5	1	2	2
6	1	2	3
7	1	3	1
8	1	3	2
9	1	3	3
10	2	1	1
11	2	1	2
12	2	1	3
13	2	2	1
14	2	2	2
15	2	2	3
16	2	3	1
17	2	3	2
18	2	3	3
19	3	1	1
20	3	1	2
21	3	1	3
22	3	2	1
23	3	2	2
24	3	2	3
25	3	3	1
26	3	3	2
27	3	3	3

noise factors. The factors having the highest noise factor can be taken as the most influential parameter in determining the response relations with the least variance. In this experiment, the lower the better criteria have been taken for wear rates and coefficient of friction. The S/N formula is given with a logarithmic dependency as follows:

$$S/N = -10\log\left[1/n\left(\Sigma y^2\right)\right] \tag{8.1}$$

Table 8.3 gives the experimental data of wear rate and coefficient of friction with different combinations of input parameters, with the preference made to the combination of parameters having the highest values. Fuzzy logic has been applied to alter the multi-response characteristic to the single response problem.

TABLE 8.3
Experimental Results and S/N Ratio

Test No.	Wear (μm)	S/N Ratio	Coefficient of Friction	S/N Ratio
1	73.987	–37.383	0.503	5.968
2	109.043	–40.752	0.42	7.535
3	92.266	–39.301	0.51	5.848
4	64.103	–36.138	0.426	7.411
5	86.827	–38.773	0.419	7.555
6	103.834	–40.327	0.441	7.111
7	91.237	–39.203	0.413	7.681
8	108.728	–40.727	0.413	7.681
9	116.154	–41.301	0.414	7.659
10	84.344	–38.521	0.568	4.913
11	87.021	–38.793	0.532	5.481
12	114.942	–41.21	0.373	8.565
13	82.285	–38.306	0.389	8.201
14	94.398	–39.499	0.452	6.897
15	107.917	–40.662	0.422	7.493
16	74.363	–37.427	0.495	6.107
17	108.631	–40.719	0.548	5.224
18	99.619	–39.967	0.478	6.411
19	65.242	–36.291	0.431	7.310
20	52.765	–34.447	0.449	6.955
21	78.045	–37.847	0.447	6.993
22	75.235	–37.528	0.39	8.178
23	82.903	–38.371	0.422	7.493
24	93.599	–39.425	0.385	8.290
25	61.305	–35.75	0.468	6.595
26	74.787	–37.477	0.388	8.223
27	92.182	–39.293	0.373	8.565

8.6 Fuzzy Logic Unit

A developed complicated system can be easily achieved using the fuzzy logic system. This system incorporates an efficient set of rules and reasoning methodology that can be described in steps of fuzzifier, an inference engine and defuzzifier. The method begins with the implication of fuzzifier with the help of a membership function (MF) in processing the S/N ratios. Then, with the help of the inference engine, the fuzzy rules and reasoning methodology are being applied to compute for a fuzzed value. Lastly, the defuzzifing activity provides a 'single MRPI' from the fuzzed values that can be used for optimizing the parameters. The two input-one output system is thus evaluated using the fuzzy technique and is shown in the following equations. The methodology is based on IF-THEN reasoning technique with two inputs taken as U_1 and U_2, and the response is taken as V.

$$\text{Rule 1: if } U_1 \text{ is } A_1 \text{ and } U_2 \text{ is } B_1, \text{then } V \text{ is } C_1 \text{ else}$$

$$\text{Rule 2: if } U_1 \text{ is } A_2 \text{ and } U_2 \text{ is } B_2, \text{then } V \text{ is } C_2 \text{ else}$$

$$\vdots$$

$$\text{Rule } n\text{: if } U_1 \text{ is } A_n \text{ and } U_2 \text{ is } B_n, \text{then } V \text{ is } C_n$$

Three MFs have been taken as $(\mu_{A_i}, \mu_{B_i}, \mu_{C_i})$, which describe the fuzzy subsets (A_i, B_i, C_i). The input to the fuzzy logic system (wear rate and coefficient of friction) has been assigned with three fuzzy subset parameters each, while the output (MRPI) has been assigned with five fuzzy subsets that are shown in Figures 8.1–8.3. By considering the highest S/N value criteria for maximum performance, 27 fuzzy rules have been applied for 27 tests. The fuzzy MF of outputs can be expressed in terms of input values of the fuzzy logic unit as U_1 and U_2, which is given as follows:

$$\mu_{C_0}(V) = \left[\mu_{A_1}(U_1) \wedge \mu_{B_2}(U_2) \wedge \mu_{C_1}(U_3) \vee \mu_{A_n}(U_1) \wedge \mu_{B_n}(U_2) \wedge \mu_{C_n}(U_3)\right] \quad (8.2)$$

where ^ is the minimum operation and ˅ is the maximum operation. The fuzzified values are then converted into non-fuzzy values by defuzzification method, which implements a centre of gravity method given as follows:

$$y_0 = \frac{\sum y\mu_{D_0}(y)}{\sum \mu_{D_0}(y)} \quad (8.3)$$

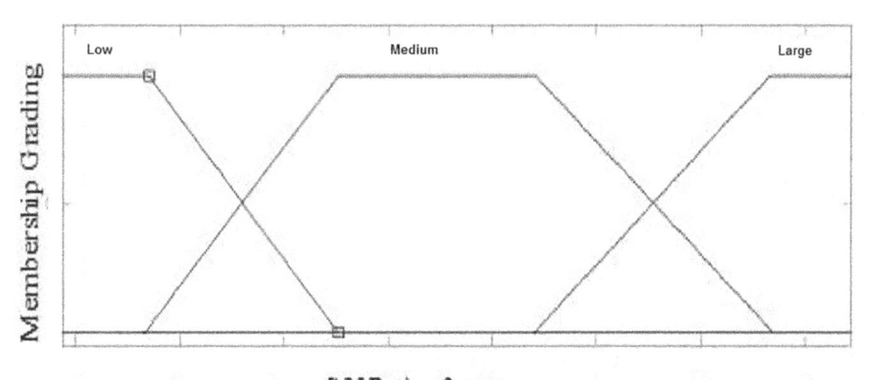

FIGURE 8.1
MF for wear.

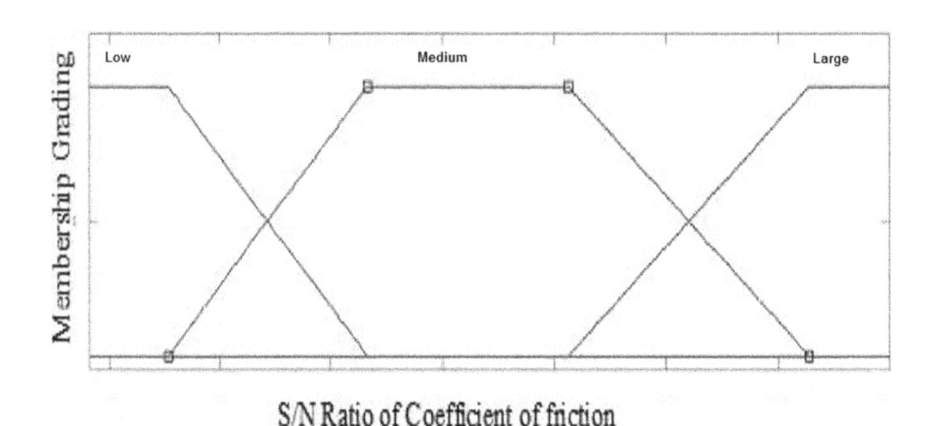

FIGURE 8.2
MF for coefficient of friction.

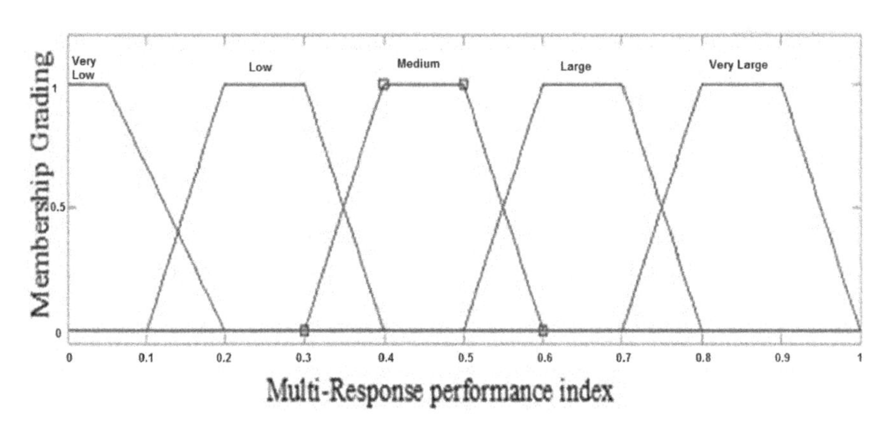

FIGURE 8.3
MF for MRPI.

The y_0 value determines the MRPI value, where a higher value indicates a better performance of the result. Table 8.4 gives the result of MRPI values for different tests.

8.7 MRPI and ANOVA for Analysis

The process is carried out by Taguchi method, and by implementing the fuzzy technique, the MRPI values have been evaluated, which facilitates a single response optimization from a multi-response. In the next step, the

TABLE 8.4
Results of MRPI

Test No.	MRPI
1	0.758
2	0.597
3	0.765
4	0.627
5	0.593
6	0.729
7	0.567
8	0.567
9	0.571
10	0.85
11	0.805
12	0.507
13	0.507
14	0.75
15	0.607
16	0.752
17	0.85
18	0.75
19	0.656
20	0.75
21	0.75
22	0.507
23	0.607
24	0.507
25	0.75
26	0.507
27	0.507

mean MRPI value is calculated from each level of parameters; likewise, three mean MRPI values are taken for the parameter in the range of (1–9), (10–18) and (19–27) in Table 8.4. Mean MRPI values at every level of each parameter are shown in Table 8.5.

MRPI denotes the effect of every parameter on the response values. The responses are checked for the mean values in MRPI at each level of process parameters. Higher MRPI value indicates higher effect of that input parameter on the output. The optimum values of parameters in determining the performance of responses can be evaluated from the response table and response graph (Figure 8.4). The optimum parameters for determining wear characteristic, considering the highest MRPI value, can be taken as parameters associated with test numbers 10 and 17 given in Table 8.4. Thus, a combination of these parameters gives a significant effect on the wear

TABLE 8.5
Response Table for MRPI

Level	A	B	C
1	3.920	3.008	3.693
2	3.146	4.468	3.609
3	4.344	3.934	4.108
Delta	1.197	1.460	0.500
Rank	2	1	3

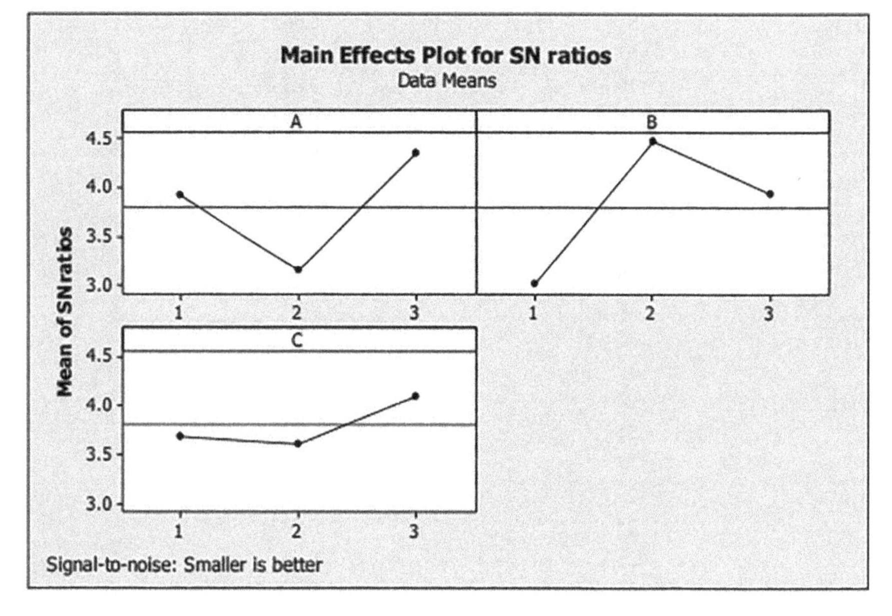

FIGURE 8.4
Main effect plot for S/N ratio for MRPI.

characteristic response. To get a more accurate effect of the parameters, the mutual interaction effect of the parameters also needs to be evaluated, which is given by ANOVA table.

ANOVA table realizes the influence of each input parameter on the output by splitting the variability in MRPI values and by taking the sum of the squares in total MRPI deviation from each parameter values.

The Fisher's value (*F* value) can also be evaluated for minimum wear and coefficient of friction (COF) characteristics to determine the maximum effective process parameters. Generally, the higher value of *F* represents greater effectiveness of process parameters on responses. Based on the ANOVA (Table 8.6), the highest influence parameter is found out to be the normal load, which has the most effect on the multi-response system

TABLE 8.6
ANOVA for MRPI

Parameter	DOF	Seq. Sum of Square (SS)	Adjusted SS	Adjusted Mean Square (MS)	F	Contribution (%)
A	2	0.041	0.041	0.020	1.71	11.81
B	2	0.056	0.056	0.028	2.35	16.13
C	2	0.007	0.007	0.003	0.29	2.01
Error	20	0.242	0.242	0.012		
Total	26	0.347				

followed by volume fraction and sliding speed. Referring the main effect plot and the response table, the optimum level in the parameters was found to be A3B2C3.

8.8 Confirmation Test

The last step is to verify the optimized parameters for the best improvement in the characterization of the wear in a silica gel based aluminium composite. Incorporating the optimum values of the parameters in MRPI, the expression can be given as follows:

$$\hat{M} = M_m + \sum_{i=1}^{n} (M_0 - M_m) \tag{8.4}$$

where M_m is the total mean of MRPI, M_0 is the mean MRPI at optimal level and n is the number of parameters that influence the multi-performance characteristics. The comparison between predicted and actual values in optimal parameter of testing is given in Table 8.7.

The validation table gives the optimal levels of parameters to be A3B2C3.

TABLE 8.7
Validation Table for MRPI

Parameter	Mean Parameter	Optimum Parameter	
		Predicted	Experimental
Combination	A2B2C2	A3B3C3	A3B2C3
S/N ratio for wear	–39.499		–39.425
S/N ratio for COF	6.898		8.291
S/N ratio for MRPI	2.499	4.624	5.900

8.9 Conclusions

1. The study was carried out based on Taguchi's optimization method, including fuzzy techniques to investigate multi-response variations with changes in combination with multi-process parameters.
2. The tests are verified and improvement in the performance of wear characteristic was found to be better with the optimal parameters selected.
3. The most significant parameter for determining wear characteristics and coefficient of friction was found to be the normal load, followed by volume fraction and sliding speed.
4. The method is cheap, systematic and a better way to evaluate the optimal process parameters.
5. The method simplifies the optimization procedure by converting the multi-response optimization to a single performance optimization, with the implementation of MRPI parameter evaluated through fuzzy rules and Taguchi techniques.

9

Multi-Objective Optimization Using Non-Dominated Sorting Genetic Algorithm

9.1 Introduction

Genetic algorithm (GA) presents an improved system of selection of input parameters for better optimization in response. It uses the principle of natural genetics. It is also utilized for searching optimal parameters and hence for optimization. Non-Dominated Sorting GA (NSGA) was under criticism due to its complexities in computation and high time consumption and poor choice of parameters by sharing the parameter σ. NSGA-II is a modified version that takes into account the drawbacks of the earlier version by eliminating the requirement of choice of a shared parameter in decision. NSGA-II namely has two main concepts: 'Population sorting' and 'Crowding distance'. Although modified yet, it preserves the elitism that defines the selection of the best solution to be carried for successive iterations. NSGA-2 is based on algorithms given as follows:

1. A 'fitness function' is associated with problem and is taken through operators, namely 'selection operator', 'crossover operator' and 'mutation operator'. After selection of the population size n, the crossover probability (p_c) and mutation probability (p_m) are defined. Then, initialize the random population size of length L and decide a feasible number of iterations t_{max} with $t_0 = 0$.
2. The strings of populations are then evaluated.
3. If the number of iterations crosses t_{max}, the activation of the termination function is initiated.
4. Strings of population are reproduced after each iteration.
5. Crossover between random strings is performed.
6. Mutation in the strings of population is performed.
7. The strings in the new populations are evaluated and taken to the next iterations in the sequence of $t = t + 1$ with the condition given in point 3, and the process is then repeated.

TABLE 9.1
Table for Input

Input Parameters (unit)	Symbols Used in the Study	Level		
		1	2	3
EC (%)	A	20	30	40
V (V)	B	12	14	16
FR (mm/min)	C	0.2	0.3	0.4
G (mm)	D	0.1	0.2	0.3

9.2 Experimental Procedure

The Taguchi method of L27 orthogonal array has been implemented in the experiment of electro-chemical machining. The input parameters considered for the study were concentration of electrolyte (EC), applied voltage (V), rate of feed (FR) and gap (G), whereas the outputs were material removal rate (MRR) and surface roughness (SR). Three levels of input parameters, as provided in Table 9.1, were utilized in the study. Tool material (copper as a cylinder with 15 mm diameter), workpiece (EN19) and electrolyte (KCl) were kept the same for all the experiments.

9.3 Experimentation

MRR which is equivalent to a rate of decrease in weight of the workpiece and SR, measured using Talysurf Tester, are taken to be the responses with maximum influence on the performance of machining. The experimental results are tabulated in Table 9.2. It can be observed that SR is minimum in experiment number 14 and is used for GA.

9.4 Mathematical Model for MRR and SR

After all measurements are done using Minitab along with the analysis of experimental data, regression equations have been formulated to establish a relationship between input parameters and output variables. The equations are as follows:

TABLE 9.2

Table for Experimentation

Exp. No.	A	B	C	D	MRR	Ra
1	1	1	1	1	27.070	3.333
2	1	1	2	2	21.550	2.500
3	1	1	3	3	24.681	2.116
4	2	1	1	2	16.560	2.020
5	2	1	2	3	21.019	2.066
6	2	1	3	1	20.621	2.399
7	3	1	1	3	28.875	6.893
8	3	1	2	1	27.283	3.970
9	3	1	3	2	29.459	2.525
10	1	2	1	2	17.649	3.861
11	1	2	2	3	18.546	2.588
12	1	2	3	1	20.170	1.606
13	2	2	1	3	20.223	2.938
14	2	2	2	1	31.848	1.569
15	2	2	3	2	33.679	1.663
16	3	2	1	1	27.726	4.278
17	3	2	2	2	31.566	2.578
18	3	2	3	3	40.491	2.353
19	1	3	1	3	24.714	2.360
20	1	3	2	1	30.255	1.754
21	1	3	3	2	28.854	2.578
22	2	3	1	1	30.175	3.696
23	2	3	2	2	31.715	3.778
24	2	3	3	3	28.530	2.781
25	3	3	1	2	27.866	3.906
26	3	3	2	3	30.891	3.191
27	3	3	3	1	47.134	1.650

$$\begin{aligned} MRR = & -3.575*(A)+4.797*(B)-8.068*(C)+1.744*(D) \\ & +1.393*(A*B)+3.425*(A*C)+1.199*(A*D) \\ & +2.251*(B*C)-3.142*(B*D)+2.031*(C*D)+20.480 \end{aligned} \tag{9.1}$$

$$\begin{aligned} SR = & 2.431*(A)+1.009*(B)-1.413*(C)-1.463*(D) \\ & -1.258*(A*B)-1.356*(A*C)+1.096*(A*D) \\ & +1.126*(B*C)+1.080*(B*D)+1.128*(C*D)+2.878 \end{aligned} \tag{9.2}$$

TABLE 9.3
Table of ANOVA

Input	Degrees of Freedom (DOF)	MRR Sum of Square (SS)	MRR Mean Square (MS)	MRR F	MRR p	Ra SS	Ra MS	Ra F	Ra p
A	2	299.811	149.906	16.991*	37.5	4.638	2.319	5.460	22.7
B	2	186.333	93.158	10.556*	23.3	0.997	0.499	1.170	4.9
C	2	129.073	64.536	7.306*	16.1	9.227	4.614	10.868	45.2
D	2	35.086	17.542	1.989	4.4	0.501	0.251	0.585	2.4
AB	4	148.251	37.063	4.199	9.3	5.757	1.439	3.393	14.1
AC	4	96.047	24.011	2.717	6.0	3.464	0.866	2.041	8.5
BC	4	55.349	13.837	1.573	3.5	0.900	0.225	0.533	2.2
Error	6	68.843	11.474			3.311	0.552		
Total	26	1018.791				28.796			

The tabulated analysis of variance (ANOVA) table (Table 9.3) has been formed to determine the suitability of the statistical model with all the output variables and results. As in Figure 9.1a, the residuals in MRR follow a normal distribution that spreads equally on both sides of the line. This proves that the regression model that is formulated provides a best fit in the parameter optimization. R^2 being 95% indicates the significance of regression model. A similar observation was also made in Figure 9.1b, which gives a probability distribution of SR (Ra) residuals, and finding the distribution pattern to be normally distributed along a straight line signifying adequacy confirms that the regression model is able to perfectly define the relation between the output variable and the input parameters.

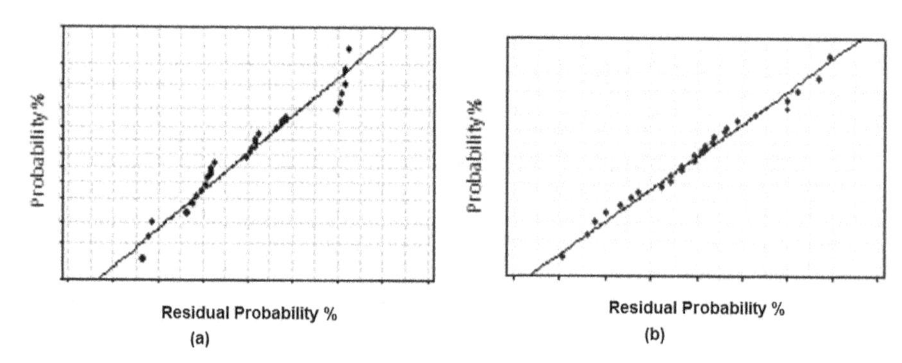

FIGURE 9.1
Plots of (a) MRR and (b) Ra.

9.5 Problem Formulation

The objective of the optimization problem in this study is to minimize the average SR and maximize the MRR. Since both the responses exhibit contradictory nature, and the GA used is minimizing in nature, a modification is required to be made to the MRR objective function to convert it to a minimized problem. The modified minimized objective function of MRR, the objective function of SR and the constraints are given as follows:

The first objective function is (MRR)^–1.

The second objective function is SR.

The boundary conditions are $A \in [20,40]; B \in [12,16]; C \in [0.2,0.4]; D \in [0.12,0.3]$.

9.6 Result and Discussion

The fitting objective functions in Section 9.5 are employed to find the optimized parameters within the limits of the upper and lower bounds and are given in Table 9.4. In Figure 9.2, a graph is plotted depicting a Pareto front between the MRR and the SR, considering the optimal input parameter setting obtained from Table 9.4. These data can be used by manufacturers, industrialists, etc. for machining. The parameters along with the levels with solution can be taken from the generated Pareto curve.

As already depicted (Table 9.2), minimum SR is obtained in the experimental run 14, with its value being 1.569 μm and the corresponding MRR value being 31.848 mm^3/min, while the corresponding input process parameters are provided in the table. Therefore, it can be inferred from Table 9.4 that the predicted SR value is within close tolerance with reference to the experimental value. The predicted SR value corresponding to the experimental run 4 in Table 9.4 is found to be in close proximity to the experimental value mentioned earlier. The SR value of 1.511 μm, the MRR value of 31.266 mm^3/min and the corresponding input parameters are selected. From this observation, it is evident that the NSGA-II solution sets are very close to the experimental data, which confirm the reproducibility of the model.

Table 9.5 presents the Validation of the proposed model vis-à-vis experimentation for optimal input parameters for MRR and SR (Ra) responses. The deviation in performance of the predicted value considering the optimal parameters and the actual value is measured and verified. An acceptable percentage error of 7.4% and 5.2% for the MRR and SR, respectively, were found from the verification table, which indicates a suitable reproducibility of the experimental conclusion.

TABLE 9.4
Pareto Results

Sl. No.	A	B	C	D	MRR	SR
1	21	9	0.35	0.33	41.513	1.483
2	19	9	0.35	0.33	39.166	1.489
3	15	12	0.33	0.33	25.118	1.529
4	15	11	0.34	0.33	31.266	1.511
5	18	10	0.34	0.33	33.286	1.504
6	15	12	0.34	0.33	27.196	1.519
7	20	9	0.35	0.33	41.540	1.482
8	11	13	0.12	0.26	21.785	1.682
9	11	13	0.20	0.33	22.444	1.596
10	13	9	0.34	0.33	33.936	1.501
11	13	12	0.34	0.33	28.125	1.516
12	15	9	0.35	0.33	38.450	1.490
13	11	11	0.34	0.33	28.229	1.516
14	13	10	0.34	0.33	34.733	1.501
15	17	9	0.34	0.33	40.440	1.486
16	11	13	0.11	0.30	21.644	1.703
17	15	11	0.34	0.33	29.109	1.513
18	13	10	0.34	0.33	35.146	1.499
19	14	12	0.32	0.32	23.629	1.544
20	11	12	0.33	0.33	24.178	1.534
21	11	11	0.34	0.33	30.689	1.511
22	13	10	0.34	0.33	36.101	1.496
23	12	13	0.17	0.32	21.979	1.638
24	11	12	0.34	0.33	25.843	1.524
25	13	13	0.28	0.32	23.064	1.562

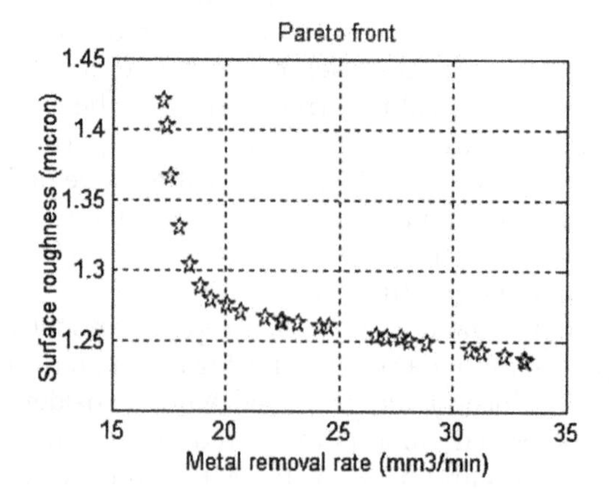

FIGURE 9.2
Pareto chart.

TABLE 9.5

Predicted Model Validation

Input				**Output**					
				MRR			**SR**		
A	**B**	**C**	**D**	**Predicted**	**Experimental**	**Error (%)**	**Predicted**	**Experimental**	**Error (%)**
30	14	0.3	0.1	31.848	31.266	1.86	1.511	1.56	3.7

9.7 Conclusion

From the earlier discussion, it is seen that a solution set is provided (Table 9.4) by implementing NSGA-2 optimization technique, which enables engineers to select the combination of parameters based on the requirement of a high MRR or a low SR in process development. The reproducibility of the relation between the responses and the process parameters is mathematically evaluated, analysed and verified using regression equation, ANOVA and confirmation test, respectively.

10

Single-Objective Optimization Using Artificial Bee Colony Algorithm

10.1 Introduction

Wire electrical discharge machining (WEDM) is an advanced machining technique employed in cutting complex and intricate-shaped workpieces with high precision and accuracy, irrespective of the nature of the material, i.e., hard material, that tends to be difficult to machine by conventional machining. It is because of the non-contact nature between the tool and the workpiece. The only constraint of this technique is that the material to be machined has to be conductive in nature. In this technique, material removal takes place due to repetitive and rapid discharge of sparks on the conductive work metal that melts and vaporizes the material. A conducting wire of diameter 0.05–0.75 mm is taken as an electrode tool that is fed along the cut path while spark discharges are kept on. The spark is caused by the discharge current that is passed through the gap between the workpiece and the wire surface filled with dielectric fluid such as de-ionized water. The purpose of the dielectric fluid is to provide current insulation across the gap, to cool down the workpiece and the tool and to flush out the eroded work material.

This chapter aims to evaluate the performance of the tool and the workpiece for the betterment of productivity. The performance is related to the surface characteristics of the machined workpiece during machining, which is influenced by the varying process parameters. Surface characteristic is determined by the average surface roughness (SR) value (Ra), which is considered a critical parameter in the variation of mechanical properties such as fatigue strength, wear resistance, wear rate, corrosion, lubrication and coefficient of friction. A minimum SR value for dimensional accuracy and surface finish was preferred and was proved to be a good predictor in characterizing the machining performance.

The chapter is mainly involved in finding the optimal combination of input parameters, viz. spark current, gap voltage, spark on-time and spark off-time for obtaining good surface characteristics in the form of minimum centre line average SR value (Ra) during the machining of EN 31 tool steel workpiece using artificial bee colony (ABC) algorithm in a WEDM process. The surface morphology is then observed using scanning electron microscopy (SEM).

10.2 Setup for Experimentation

The experiments are carried out on a computer numerical control (CNC)-type WEDM, and de-ionized water is used as dielectric fluid. The workpiece and the tool are taken as EN 31 tool steel and 0.25-diameter zinc-coated brass wire, respectively. As already stated, machining parameters considered are spark current, gap voltage, spark on-time and spark off-time, which are expected to have maximum influence on the SR of the finished workpiece. Table 10.1 provides the parameters and their different considered values. The table represents the design parameters where the other influential parameters are taken to be constant or having negligible effect on the responses. Central composite design (CCD) methodology with 31 experiments is used for evaluating SR (Table 10.2). A profilometer is used to measure SR.

10.3 Result and Discussion

After the assessment of SR response values influenced by machining parameters, a model is generated providing the cause-and-effect analysis of response values and machining parameters using response surface methodology (RSM). The empirical relation is presented as follows:

TABLE 10.1

Tabulation of Input Parameters with Different Values

			CCD Levels				
Parameters	**Unit**	**Symbol**	**−2**	**−1**	**0**	**1**	**2**
Spark current	A	A	3	5	7	9	11
Gap voltage	V	B	20	30	40	50	60
Spark on-time	μs	C	3	4	5	6	7
Spark off-time	μs	D	1.5	3	4.5	6	7.5

TABLE 10.2

Experimental Results

Exp. No.	A	B	C	D	Ra
1	–1	–1	0	0	14.98
2	1	–1	–1	–1	13.21
3	–1	–1	–1	1	12.53
4	0	0	0	0	13.69
5	0	0	0	–2	15.25
6	0	2	0	0	13.69
7	1	–1	1	–1	15.15
8	–1	–1	–1	–1	11.89
9	1	–1	–1	1	14.14
10	0	0	0	0	13.69
11	–1	1	1	1	14.55
12	1	0	0	0	14.13
13	1	1	1	1	16.23
14	–1	1	–1	–1	11.99
15	0	0	0	0	13.69
16	0	0	0	0	13.69
17	2	0	0	0	14.60
18	0	0	0	2	14.23
19	0	0	2	0	16.08
20	0	–2	0	0	14.93
21	1	1	–1	1	12.24
22	0	0	–2	0	8.16
23	0	0	0	0	13.69
24	1	1	1	–1	15.43
25	0	0	0	0	13.69
26	–1	1	–1	1	11.55
27	–2	0	0	0	11.91
28	0	0	0	0	13.69
29	–1	–1	1	–1	14.78
30	1	–1	1	1	15.75
31	–1	1	1	–1	14.50

$$
\begin{aligned}
R_a = & -0.25*(A)+0.24*(B)+1.40*(C)-0.33*(D) \\
& -0.002*(A)^2-0.002*(B)^2-0.28*(C)^2-0.07*(D)^2 \\
& +0.01*(A*B)+0.03*(A*C)+0.03*(A*D) \\
& -0.01*(B*C)-0.01*(B*D)+0.08*(C*D)-0.35
\end{aligned} \tag{10.1}
$$

A code for ABC algorithm is generated in MATLAB® based on the earlier relation with the following boundary conditions. The selection of this particular condition has been devised for proper convergence of the optimization problem.

If the number of population = ***n***;

Then, the number of employed bees = number of onlooker bees = **0.5*n**

And the number of scout bees per cycle = **0.1*n**.

In this case, ***n*** is taken as 10 with 1,000 cycles with a limit of 500.

As solutions are taken through 1,000 iterations, an optimal solution is obtained at the end of the iteration process. Figure 10.1 depicts the variation in SR with the number of iterations. From this figure, it is concluded that the best optimal solution is achieved at the end of the iteration that yields the maximum value of the result where no further improvements are possible. Figure 10.2a and b, respectively, shows the variation of SR with respect to four WEDM process parameters. From the figure, it is clear that the SR value decreases with decrease in the current and pulse off-time, while it increases

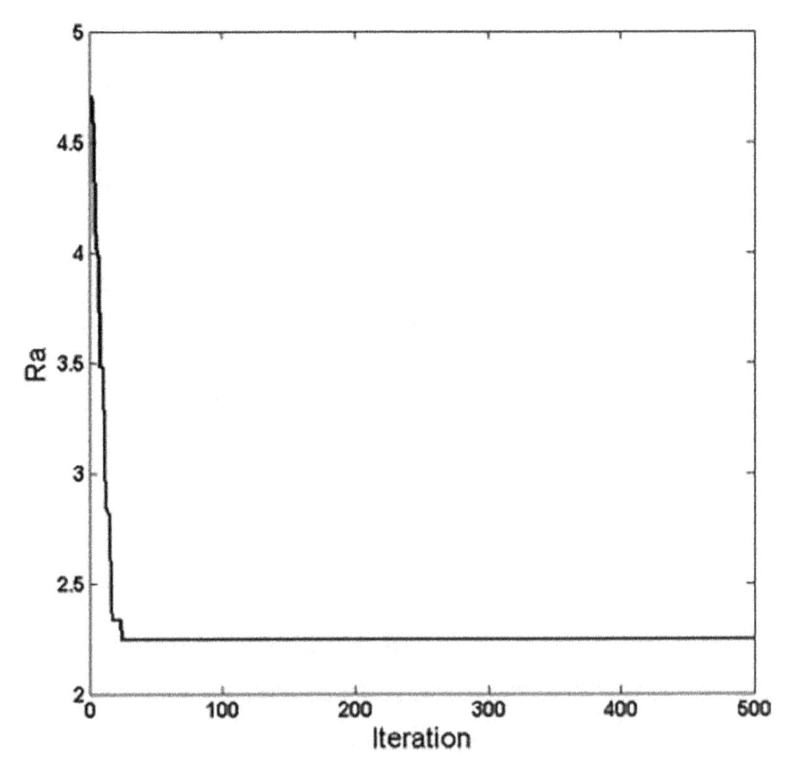

FIGURE 10.1
Convergence graph.

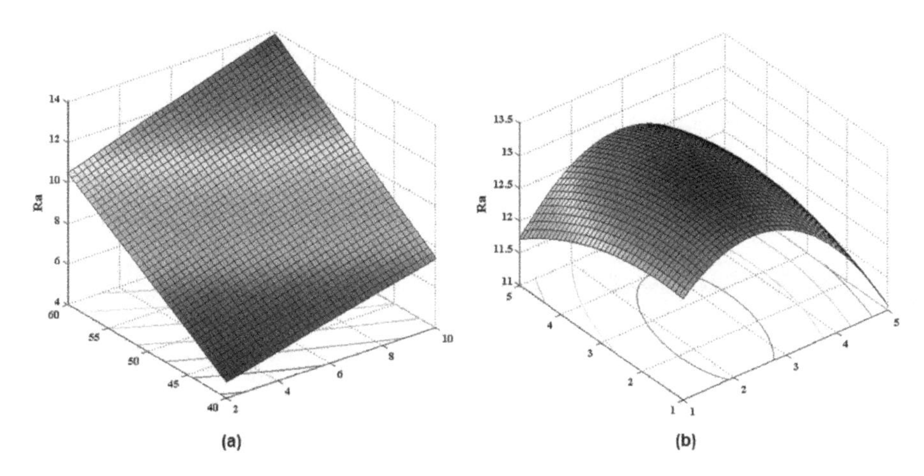

FIGURE 10.2
Contour plots for SR: (a) spark current and gap voltage; (b) spark on-time and spark off-time.

with decrease in pulse on-time. Thus, the analysis presents optimal values of combination of parameters within a specified range, and the values are given as follows: spark current = 3 A, gap voltage = 60 V, spark on-time = 7 μs and spark off-time = 7.5 μs. These values are set for a minimum value of SR which is evaluated to be 14.31 μm.

An experiment is then performed with all the optimal parameters being taken as given earlier, and the SR value is found to be 14.24 μm. Thus, the error between the predicted and experimental values in SR is found to be 0.49%, which proves the suitability of the ABC algorithm-based optimization model in providing a better result for SR.

10.4 Surface Morphology Analysis

Surface morphology is studied using SEM by analysing the microstructure of the workpiece before machining, while machining at the mean values of the input parameters and after machining (Figure 10.3). The surface of the workpiece before the machining is observed to be smooth with no globular spot, while it is much rougher having unevenly distributed globules on its machined surface after machining. Due to the intense heat of the spark discharges, the work material melts and evaporates, while the dielectric liquid flushes out the eroded material and rapidly cools down the melted material. This rapid cooling of material results in high thermal stresses, causing large size craters, hollow cavities and micro-cracks on the machined surface.

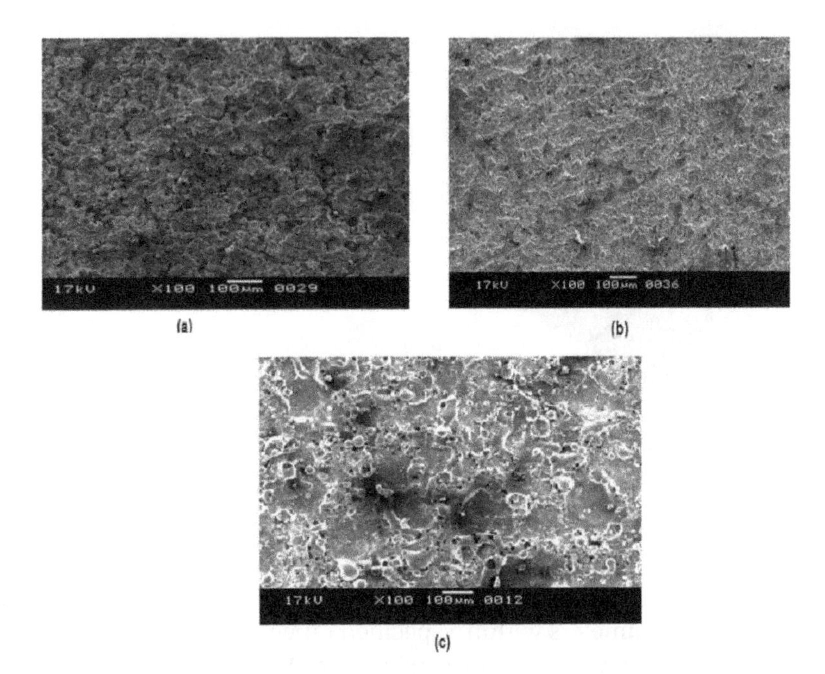

FIGURE 10.3
SEM micrographs (a) before machining, (b) while machining at the mean values and (c) after machining.

10.5 Conclusion

The study aims at evaluating the optimal parameters (spark current, gap voltage, spark on-time and spark off-time) in WEDM for a minimum SR on the machined surface of a workpiece by applying the ABC algorithm. The design of the experiment was based on the CCD (rotatable), and RSM was applied for the evaluation of regression equation for SR. For minimal SR, the optimal combination of parameter values is given as follows: spark current = 3 A, gap voltage = 60 V, spark on-time = 7 µs and spark off-time = 7.5 µs. A confirmation test was performed to conform the predicted result to the experimental results, and the error between them was found to be small (<1%). From the 3D contour plot, the SR parameter was found to be proportional to spark current, gap voltage and spark on-time, and to identify the surface morphological character before and after machining, an SEM study was performed.

Section III

Hands-On Training on Various Software Dedicated for the Usage of Techniques

11

Working with Minitab for Optimization Using Taguchi Technique

11.1 Introduction

Hands-on training on Minitab, a statistical software, to perform a single-objective optimization using Taguchi design for a given problem.

11.2 Problem Definition

To optimize the performance of a chemical reaction under various controlled factors such as *temperature, pressure, concentration* and *surface area. Heat release* is taken as the output for the process.

The various levels of controlled parameters are shown in Table 11.1;

11.3 Software Operation

Step 1: To create a Taguchi design

Stat > DOE > Taguchi > Create Taguchi Design (Figure 11.1).

TABLE 11.1

Levels of Process Parameters

		Levels		
Process Parameter (unit)	**Symbol**	**1**	**2**	**3**
Temperature	A	20	30	40
Pressure	B	16	20	24
Concentration	C	0.20	0.41	0.62
Surface area	D	0.4	0.5	0.6

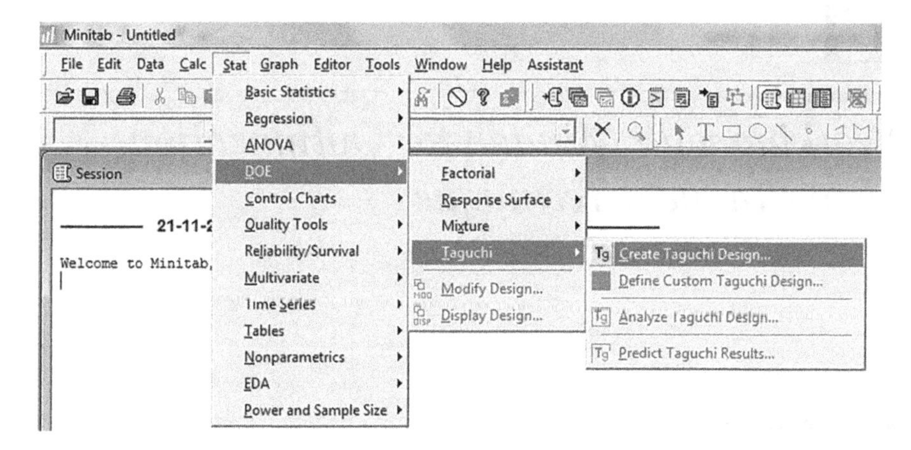

FIGURE 11.1
Initial call-out.

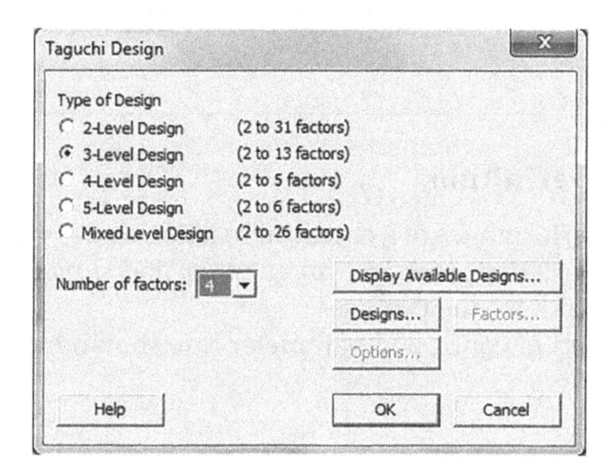

FIGURE 11.2
Design window.

This would open a new window shown in Figure 11.2. Here, select the level of design parameter and the number of factors entitled in the problem. Four controlled factors and three levels for each are taken here as problem definition, and click the Designs button, a small window appearing on the window.

It would generate another call-out for selecting the Design of Experiments (Figure 11.3).

The window would provide two options, namely L9 and L27. Since the interaction is expected to occur, L27 orthogonal array for experiment is chosen. Then press OK.

The two call-outs generated are shown in Figure 11.4. Here,

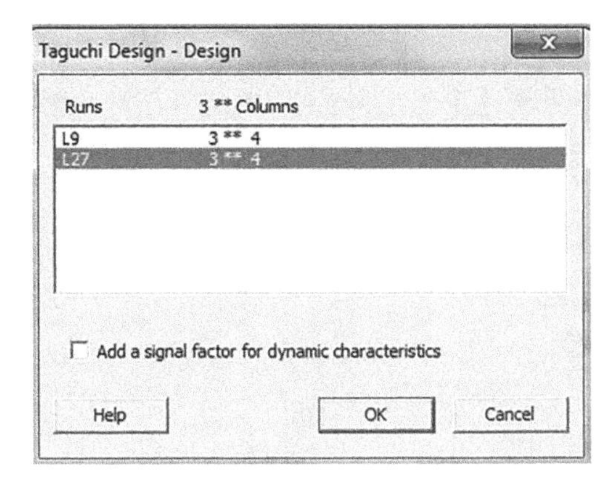

FIGURE 11.3
Window for design of experiments (DOE).

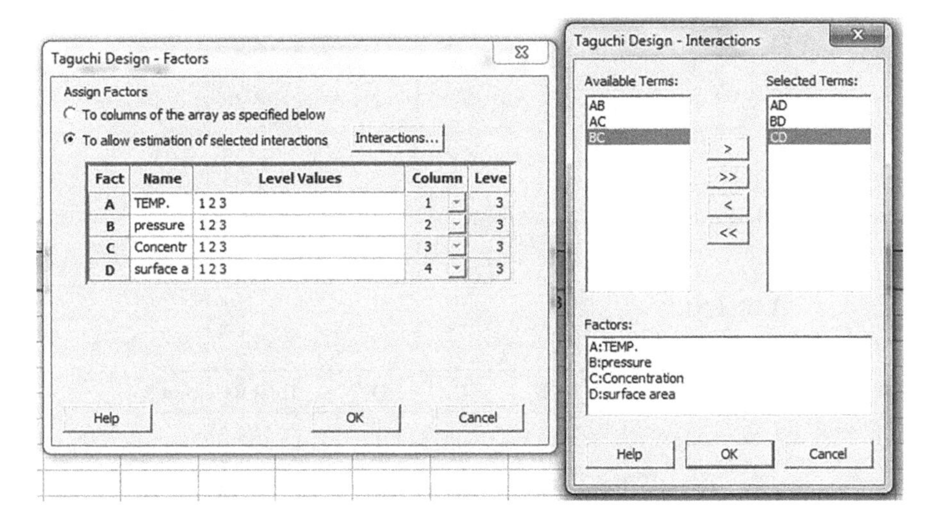

FIGURE 11.4
Window for interactions DOE.

- Step 1.2: Selection of interaction

  Factors > Interactions > select the suitable interaction> OK > OK > OK.

  This generates an orthogonal array for conducting the experiment (27 experiments for L27) (Figure 11.5)

  The experiment is conducted by varying the parameters, and the result is recorded as shown in Table 11.2.

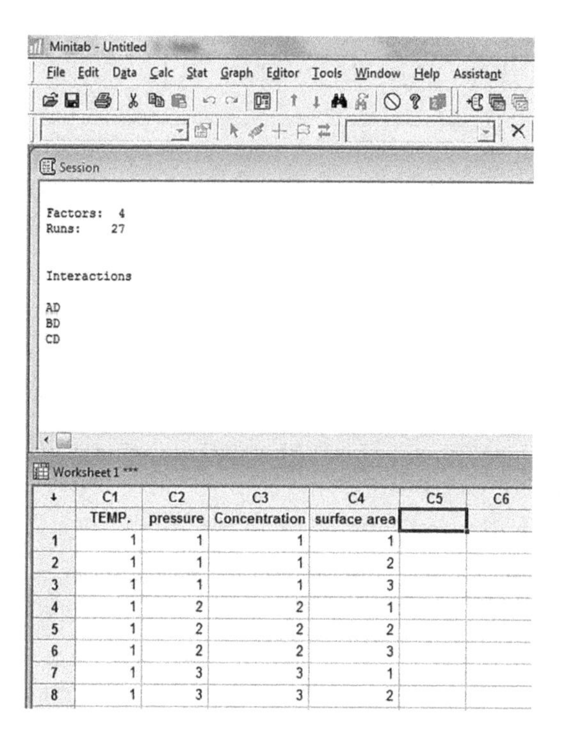

FIGURE 11.5
Window for orthogonal array (OA) generation.

TABLE 11.2
Experimental Results

Exp. No.	A	B	C	D	Heat Released
1	1	1	1	1	52
2	1	1	2	2	42
3	1	1	3	3	35
4	1	2	1	2	45
5	1	2	2	3	27
6	1	2	3	1	69
7	1	3	1	3	21
8	1	3	2	1	62
9	1	3	3	2	86
10	2	1	1	2	48
11	2	1	2	3	51
12	2	1	3	1	32
13	2	2	1	3	32
14	2	2	2	1	44
15	2	2	3	2	57

(Continued)

TABLE 11.2 (*Continued*)

Experimental Results

Exp. No.	A	B	C	D	Heat Released
16	2	3	1	1	70
17	2	3	2	2	83
18	2	3	3	3	33
19	3	1	1	3	46
20	3	1	2	1	60
21	3	1	3	2	64
22	3	2	1	1	33
23	3	2	2	2	66
24	3	2	3	3	82
25	3	3	1	2	20
26	3	3	2	3	55
27	3	3	3	1	76

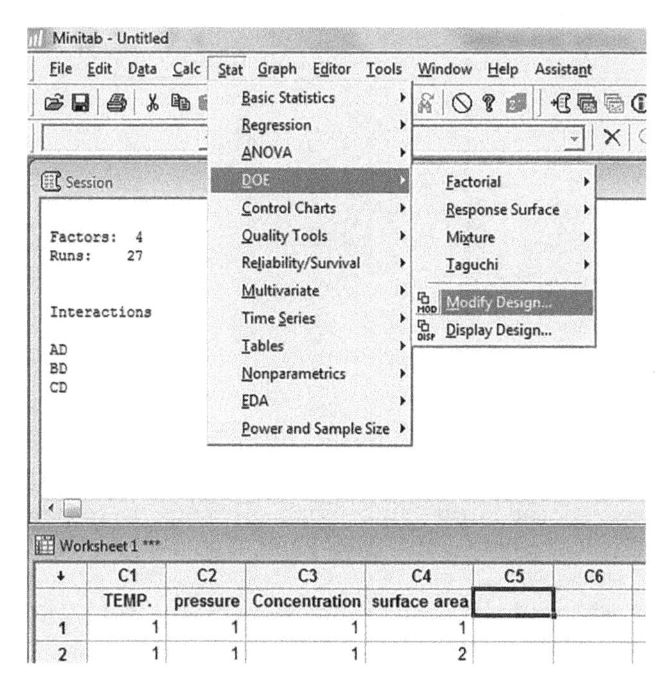

FIGURE 11.6
Design modification window.

Step 2: Modifying the orthogonal array
Stat > DOE > Modify Design > OK (Figure 11.6)

- 2.2 Input level value
  Enter the level of each input as mentioned in the problem, as indicated in Figure 11.7.

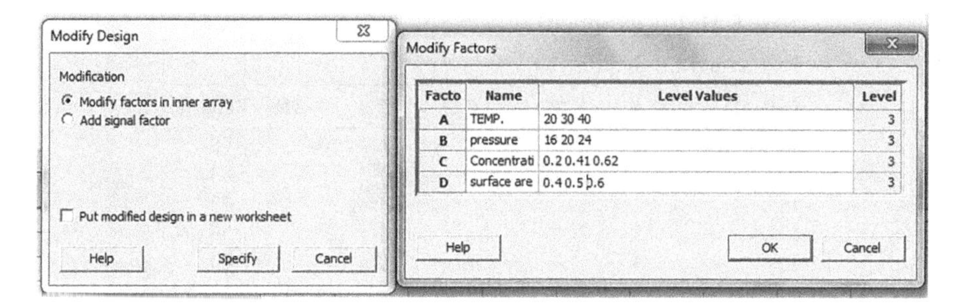

FIGURE 11.7
Input window.

Worksheet 1 ***

↓	C1	C2	C3	C4	C5	C6	C
	TEMP.	pressure	Concentration	surface area	Heat release		
1	20	16	0.20	0.4	52		
2	20	16	0.20	0.5	41		
3	20	16	0.20	0.6	38		
4	20	20	0.41	0.4	45		
5	20	20	0.41	0.5	27		
6	20	20	0.41	0.6	69		
7	20	24	0.62	0.4	21		

FIGURE 11.8
Response window.

Step 3: Input the response in the orthogonal array

This can be done by entering one input similar to writing in an Excel sheet or can be directly copying or pasting the response. After entering the response, the table looks like the one as shown in Figure 11.8.

Now we input target and other options step by step as follows (Figure 11.9):

- 3.1 Stat > DOE > Taguchi > Analyze Taguchi Design
- 3.2 Select the response 'Heat release'.
- 3.3 Click 'Options'.

  Select 'Smaller is better' as here 'Heat release' is to be minimized, and press OK (Figure 11.10).
- 3.4 Click 'Storage'.

  Tick the required results; here, Signal-to-Noise ratios is selected (Figure 11.11).

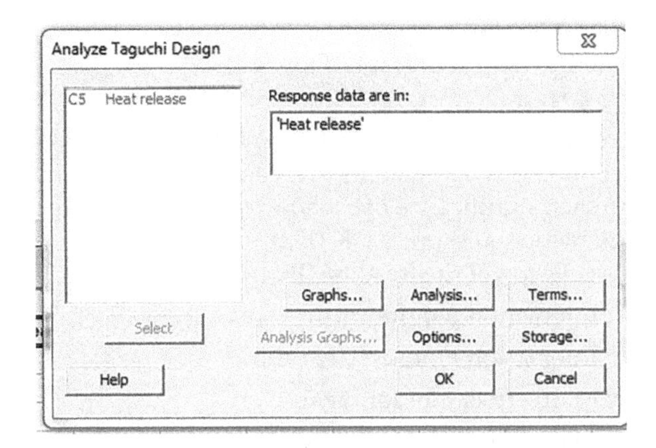

FIGURE 11.9
Analysis selection window.

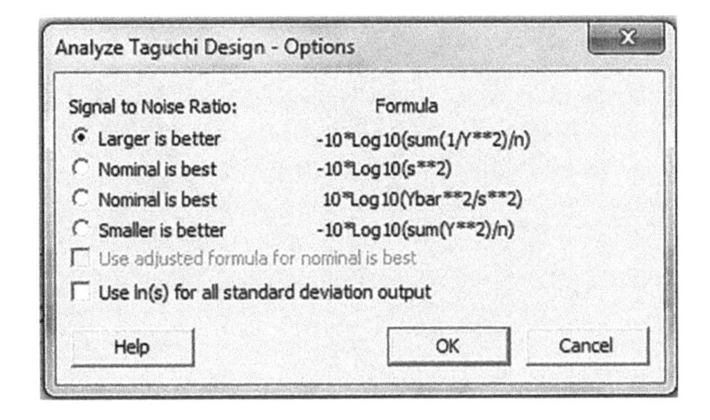

FIGURE 11.10
Criterion selection window.

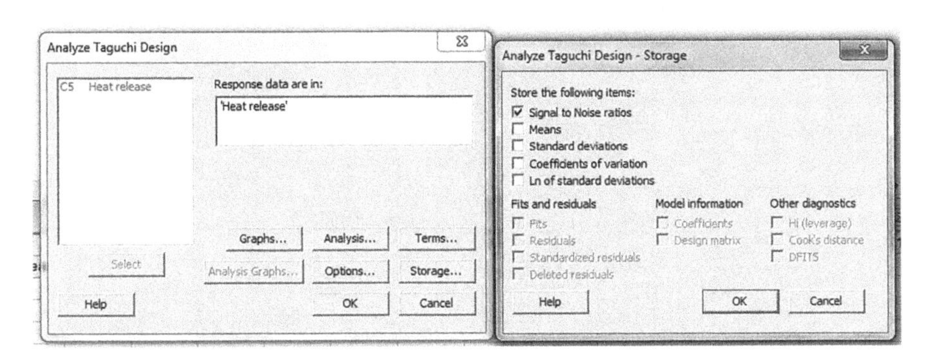

FIGURE 11.11
Resultrequirement window.

- 3.5 Click 'Graphs'.
  Select the required graph (Figure 11.12).
- 3.6 Click 'Terms'.
  Here, interactions are selected, and the maximum of the three interactions can be taken in this case. This selection is based on the impact/significance of each factor. Use arrow button to select the interaction and press OK (Figure 11.13).
  Press OK to get the 'results' (Figures 11.14 and 11.15).

Plot 1: Interaction plot for mean.

Plot 2: Interaction plot for S/N ratio.

Plot 3: Main effect plot for mean.

Plot 4: Main effect plot of S/N ratio.

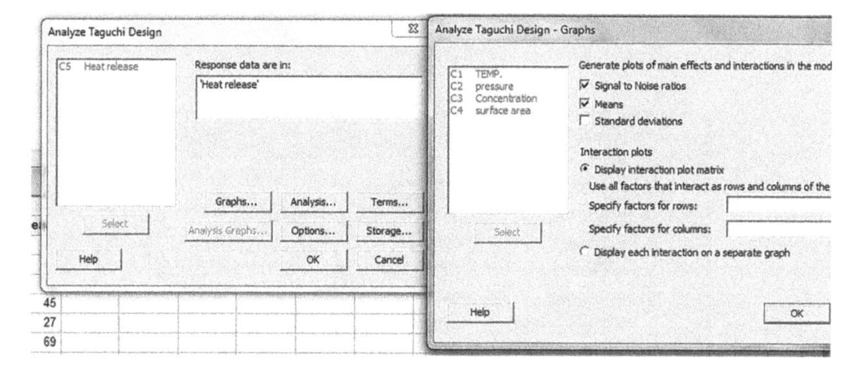

FIGURE 11.12
Analysis window.

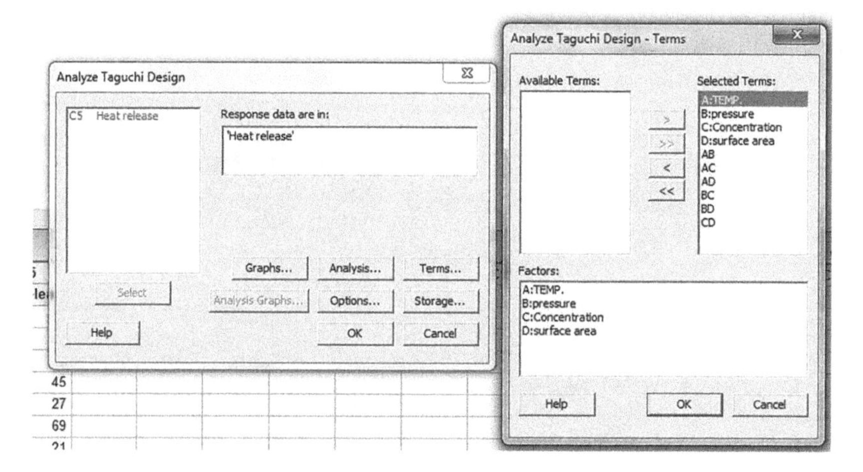

FIGURE 11.13
Confirmation window.

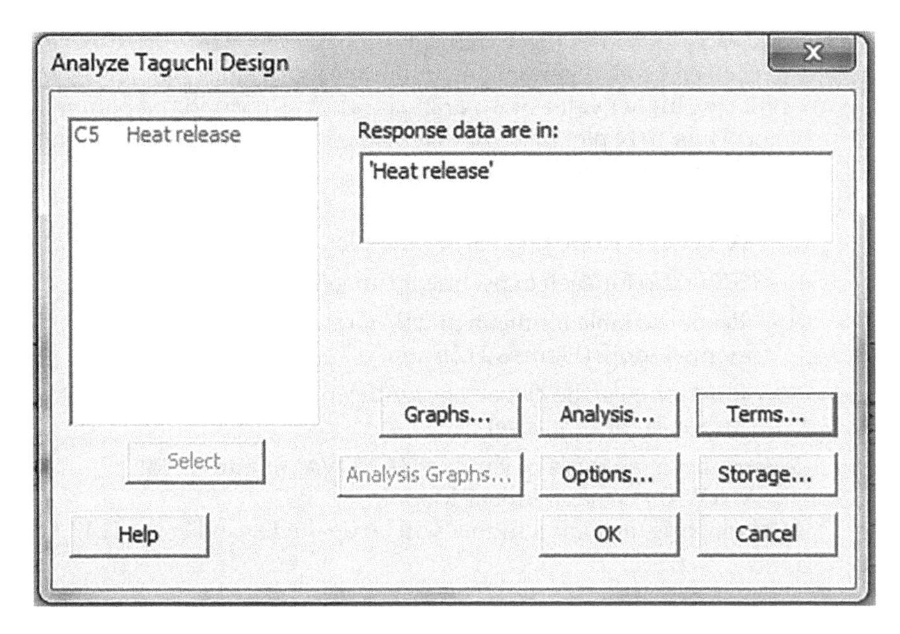

FIGURE 11.14
Result confirmation window.

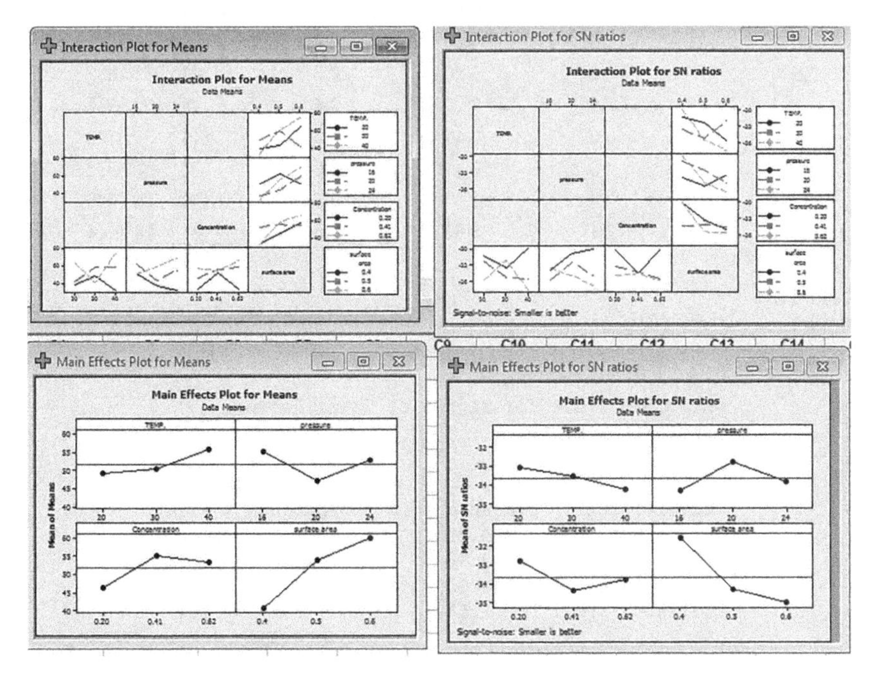

FIGURE 11.15
Interaction plots.

Mean plot shows the actual variation of response with respect to each controlled parameter, and in the main effect plot for S/N ratio, the higher value of S/N ratio is taken as the optimal parameter. Thus, S/N plot helps to determine the optimal combination for the aforementioned problem.

Results:

- A1B2C1D1 from mean S/N ratio plot.
- S/N ratio for each experiment run gets stored in worksheet.
- Response table for mean and S/N ratio gets stored in the session window (Figures 11.16 and 11.17).
- Rank shows the 'level of significance' of each factor over response (heat release).

Step 4. Performing analysis of variance (ANOVA) (Figure 11.18)

Stat > ANOVA > General Linear Model

After clicking, a small window will be opened as in Figure 11.19.

Worksheet 1 ***

↓	C1	C2	C3	C4	C5	C6
	TEMP.	pressure	Concentration	surface area	Heat release	SNRA1
1	20	16	0.20	0.4	52	-34.3201
2	20	16	0.20	0.5	41	-32.2557
3	20	16	0.20	0.6	38	-31.5957
4	20	20	0.41	0.4	45	-33.0643
5	20	20	0.41	0.5	27	-28.6273
6	20	20	0.41	0.6	69	-36.7770
7	20	24	0.62	0.4	21	-26.4444

FIGURE 11.16
S/N ratios.

Response Table for Signal to Noise Ratios
Smaller is better

Level	TEMP.	pressure	Concentration	surface area
1	-33.07	-34.34	-32.79	-31.57
2	-33.57	-32.76	-34.36	-34.31
3	-34.28	-33.82	-33.77	-35.03
Delta	1.21	1.57	1.57	3.47
Rank	4	2	3	1

FIGURE 11.17
Response table.

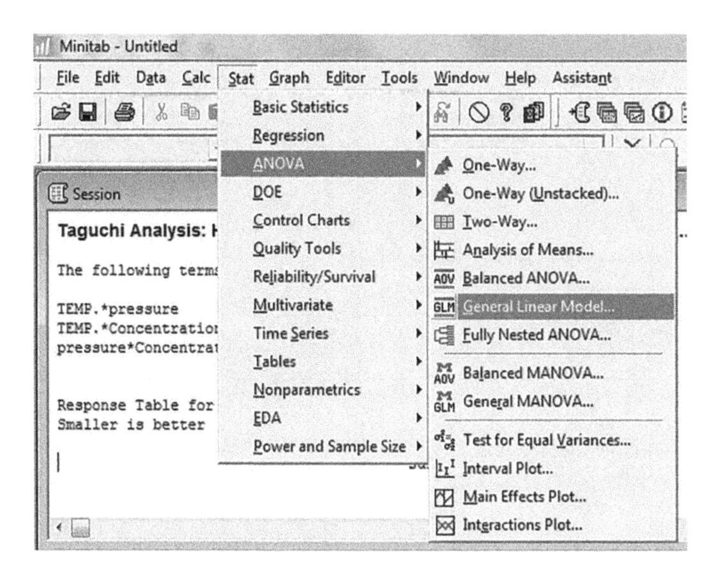

FIGURE 11.18
ANOVA table.

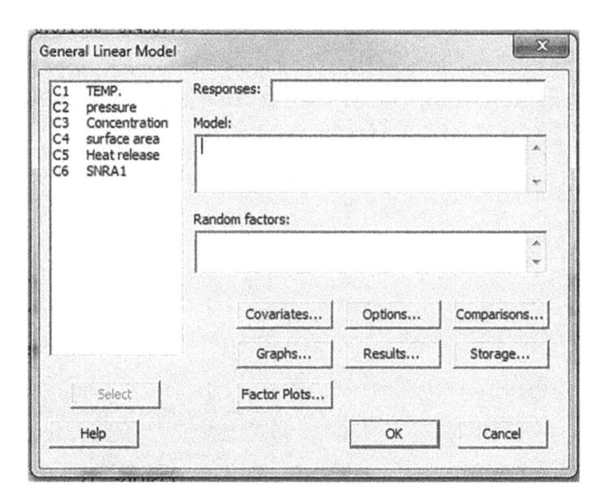

FIGURE 11.19
Input window for ANOVA.

Fill response and models as shown in the figure and click OK. ANOVA table appears in the session window (Figures 11.20 and 11.21).

Step 5. Regression analysis

Stat > Regression > General Regression > OK

As the window appears on the screen, as shown in Figure 11.22, fill the response and model. Then, click 'Result'. Select all the required

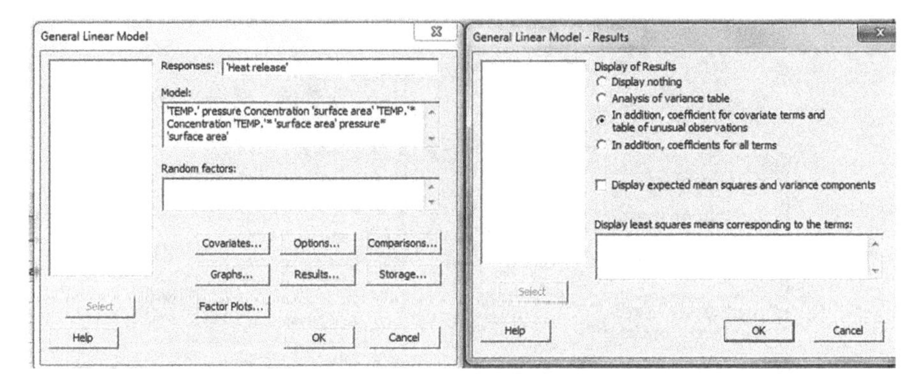

FIGURE 11.20
Second input window for ANOVA.

Analysis of Variance

Source	DF	Seq SS	Adj SS	Adj MS	F	P
Regression	7	2428.33	2428.33	346.905	0.952306	0.491580
TEMP.	1	213.56	6.21	6.207	0.017038	0.897519
pressure	1	32.00	2.08	2.081	0.005713	0.940538
Concentration	1	220.50	52.03	52.032	0.142835	0.709668
surface area	1	1682.00	9.97	9.966	0.027357	0.870377
TEMP.*Concentration	1	58.80	31.11	31.111	0.085405	0.773270
TEMP.*pressure	1	13.14	13.14	13.144	0.036083	0.851358
TEMP.*surface area	1	208.33	208.33	208.333	0.571906	0.458777
Error	19	6921.30	6921.30	364.279		
Total	26	9349.63				

FIGURE 11.21
ANOVA results.

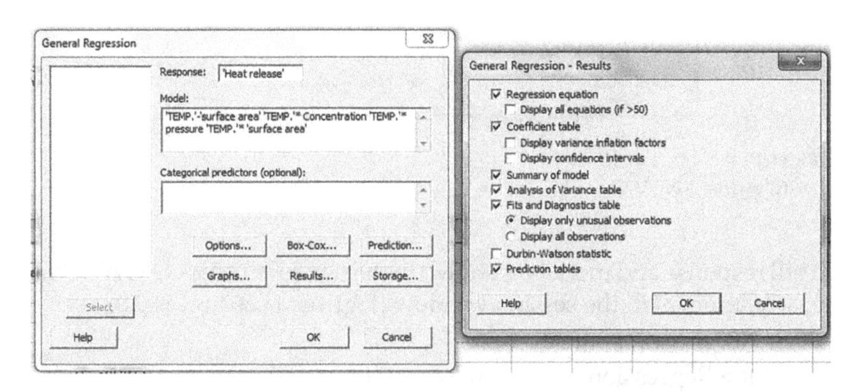

FIGURE 11.22
Input for regression equation.

terms and equations and click OK to generate the regression equation and store it in the session window.

Result: Regression result is obtained (Figure 11.23).

Step 6: Find the predicted value (Figure 11.24).

Stat >DOE > Taguchi >Predict Taguchi Results

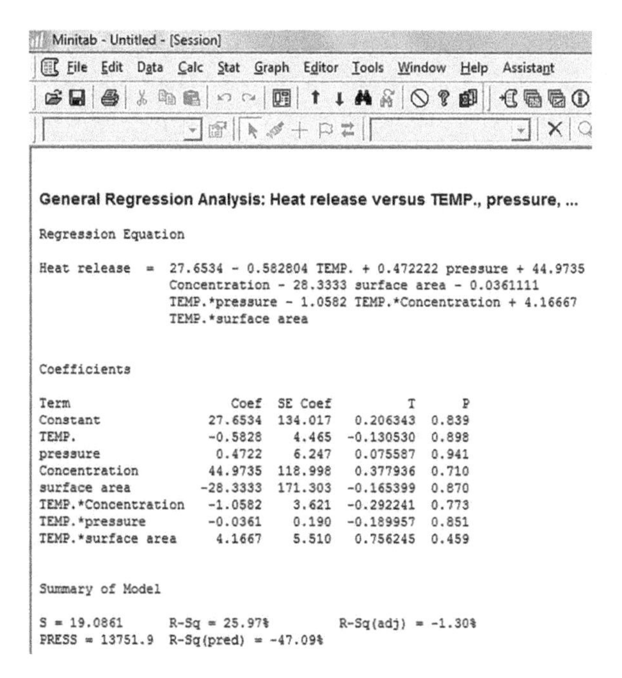

General Regression Analysis: Heat release versus TEMP., pressure, ...

Regression Equation

Heat release = 27.6534 - 0.582804 TEMP. + 0.472222 pressure + 44.9735 Concentration - 28.3333 surface area - 0.0361111 TEMP.*pressure - 1.0582 TEMP.*Concentration + 4.16667 TEMP.*surface area

Coefficients

Term	Coef	SE Coef	T	P
Constant	27.6534	134.017	0.206343	0.839
TEMP.	-0.5828	4.465	-0.130530	0.898
pressure	0.4722	6.247	0.075587	0.941
Concentration	44.9735	118.998	0.377936	0.710
surface area	-28.3333	171.303	-0.165399	0.870
TEMP.*Concentration	-1.0582	3.621	-0.292241	0.773
TEMP.*pressure	-0.0361	0.190	-0.189957	0.851
TEMP.*surface area	4.1667	5.510	0.756245	0.459

Summary of Model

S = 19.0861 R-Sq = 25.97% R-Sq(adj) = -1.30%
PRESS = 13751.9 R-Sq(pred) = -47.09%

FIGURE 11.23
Regression equation.

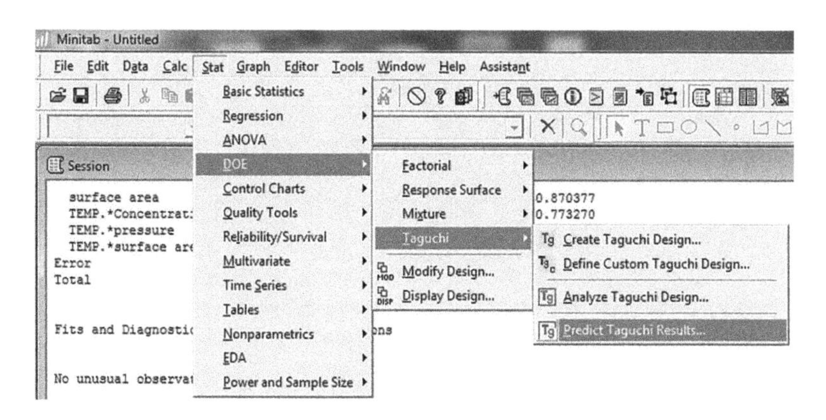

FIGURE 11.24
Window for result prediction.

Select the level of control factor and tick the required result. Click OK to store the result in the session window shown in Figure 11.25. Predicted result is finally provided in Figure 11.26.

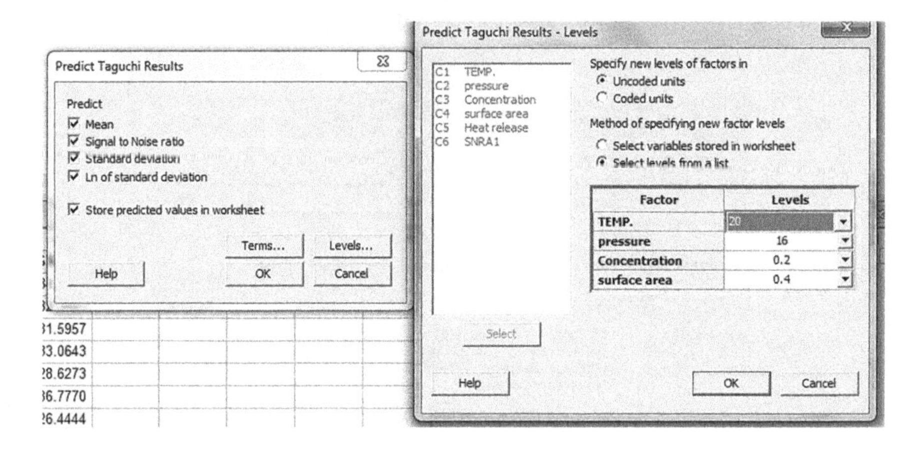

FIGURE 11.25
Input window for result prediction.

```
S/N Ratio      Mean
 -31.8756  42.8519


Factor levels for predictions

                                         surface
TEMP.  pressure  Concentration              area
   20        16            0.2               0.4
```

FIGURE 11.26
Predicted results.

12

Working with MATLAB® for Optimization Using Artificial Neural Network

12.1 Introduction

Hands-on training on MATLAB®, a statistical software, is used to perform multi-objective optimization using artificial neural network (ANN) for a given problem.

12.2 Problem Definition

To optimize the performance of a chemical reaction under various controlled factors such as *temperature, pressure, concentration* and *surface area. Heat and CO_2 release* is taken as the output for the process.

The various levels of controlled parameters are shown in Table 12.1.

Experiments are conducted by varying parameters, and the result is recorded as shown in the table (Table 12.2) (same as Taguchi design of experiments).

TABLE 12.1

Levels of Process Parameters

		Levels		
Process Parameter (unit)	**Symbol**	**1**	**2**	**3**
Temperature	A	20	30	40
Pressure	B	16	20	24
Concentration	C	0.20	0.41	0.62
Surface area	D	0.4	0.5	0.6

TABLE 12.2

Experimentation Table with Results

Exp. No.	A	B	C	D	Heat	CO_2
1	1	1	1	1	52	50
2	1	1	2	2	42	62
3	1	1	3	3	35	64
4	1	2	1	2	45	59
5	1	2	2	3	27	81
6	1	2	3	1	69	30
7	1	3	1	3	21	83
8	1	3	2	1	62	41
9	1	3	3	2	86	18
10	2	1	1	2	48	54
11	2	1	2	3	51	51
12	2	1	3	1	32	80
13	2	2	1	3	32	74
14	2	2	2	1	44	61
15	2	2	3	2	57	45
16	2	3	1	1	70	30
17	2	3	2	2	83	24
18	2	3	3	3	33	67
19	3	1	1	3	46	56
20	3	1	2	1	60	44
21	3	1	3	2	64	39
22	3	2	1	1	33	68
23	3	2	2	2	66	32
24	3	2	3	3	82	29
25	3	3	1	2	20	86
26	3	3	2	3	55	47
27	3	3	3	1	76	29

12.3 Software Operation

Step 1: Load input and output data in MATLAB workspace.

Step 2: Select a suitable neural network for a given data.

- Import data set in the Neural Network/Data Manager window.
- Select proper algorithms, learning rate and transfer function.

Step 3: Train the network and validate the test result.

- Minimize mean square error (MSE) and make regression correlation (RC) close to one (RC = 1, ideal).

Step 4: Simulate the train network for the desired input and find the predicted value.

12.4 Details of the Operations

Step 1: Load data sets in MATLAB workspace

Create a spreadsheet/matrix in MATLAB workspace and name as input, output and sample, as shown in Figure 12.1.

Click the input sheet and insert the input result only, as shown in Figures 12.2 and 12.3.

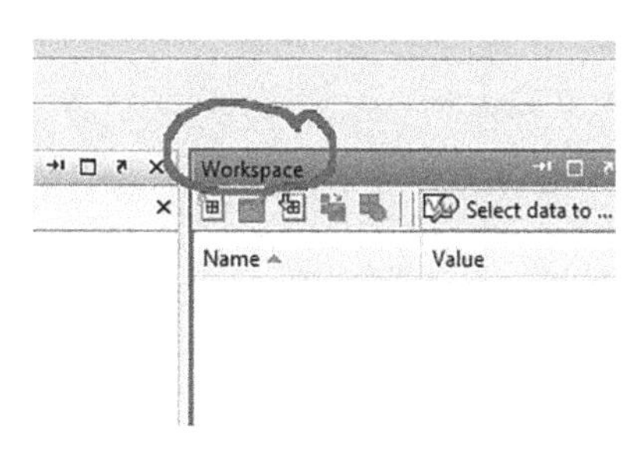

FIGURE 12.1
MATLAB workspace.

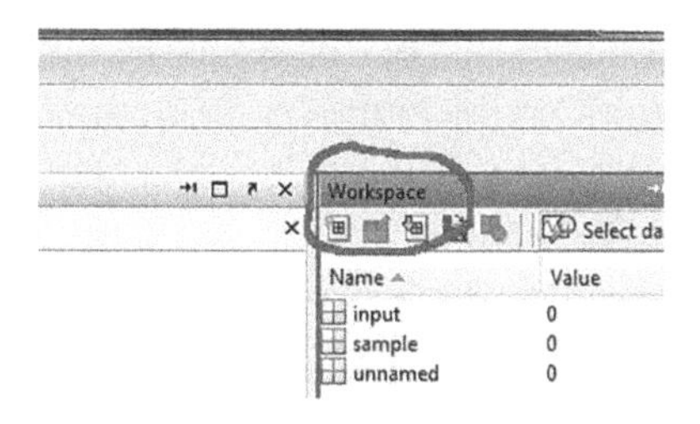

FIGURE 12.2
Inserting data in a spreadsheet/matrix.

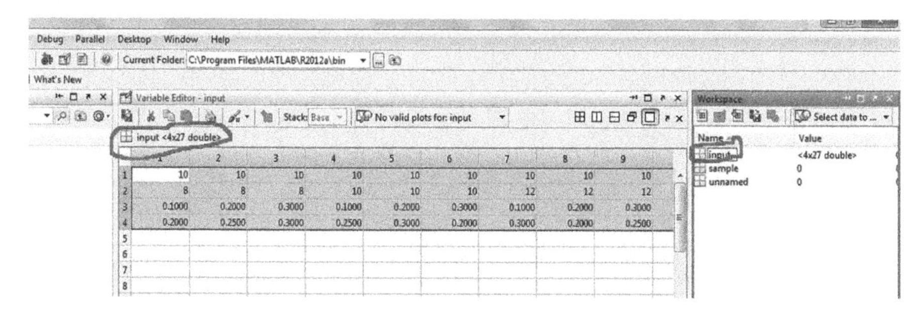

FIGURE 12.3
Inserting data in the matrix generated.

Similarly, insert data for output and sample data (here input data is taken as sample data). Here, two responses are considered (heat release and CO_2 release) (Figure 12.4).

Step 2: Select a suitable neural network

Type 'nntool' in the 'command window'.

>>nntool

This gives an application window as shown in Figure 12.5.

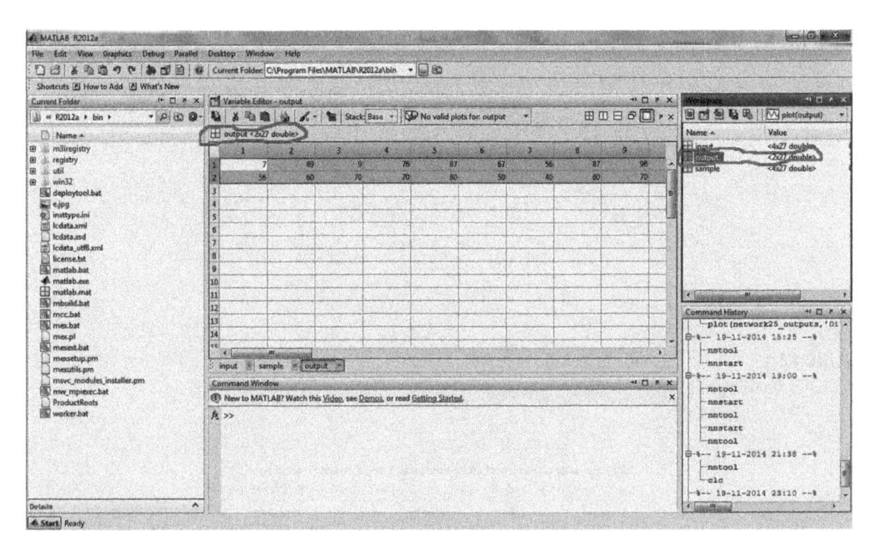

FIGURE 12.4
Inserting responses.

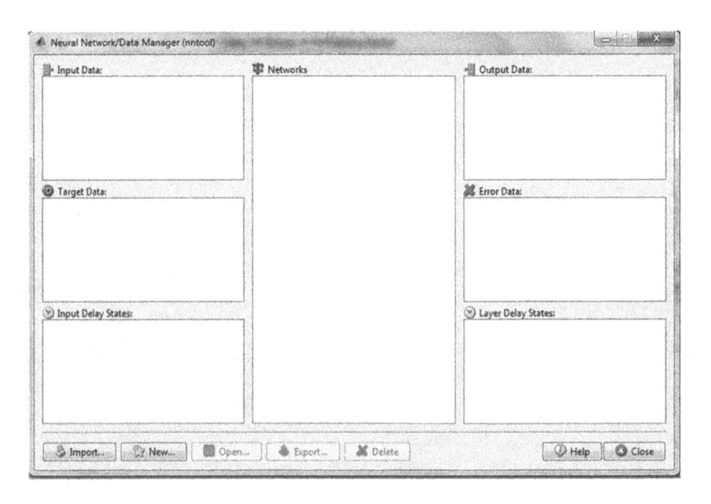

FIGURE 12.5
Application widow.

Import data set in the Neural Network/Data Manager window

Click the 'Import' button shown in the Neural Network/Data Manager window, which gives a window as shown in Figure 12.6.

Now import the 'input data' set => input data, import the 'output data' set => target data and close this window.

Creating neural network

Now click the 'New' button in the Neural Network/Data Manager window to create a new ANN model. It gives a window as shown in Figure 12.7.

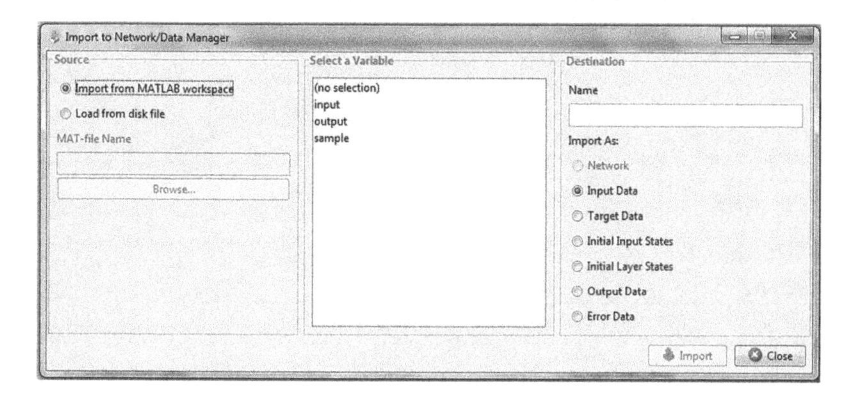

FIGURE 12.6
Import operation.

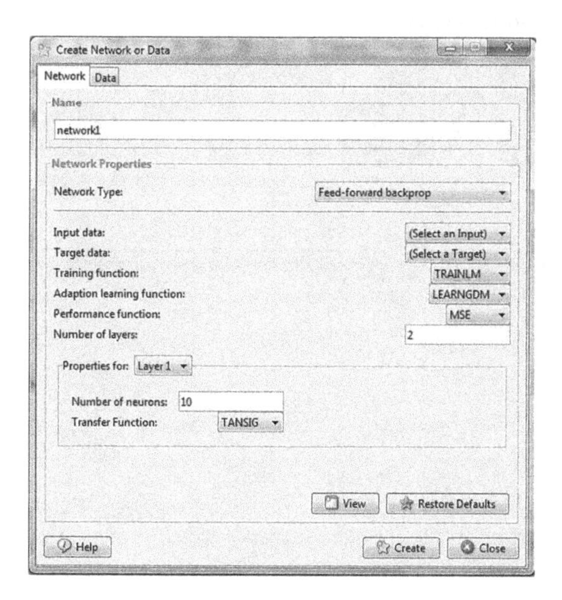

FIGURE 12.7
Creating neural network.

Select

1. Network type
2. Input
3. Target (output value)

All the other required functions such as training function, performance function and number of neurons, and click 'Create'.

The neural network can be viewed by clicking the 'View' button. Now, close this window.

A neural network generated for a given data set is shown in the view mode (Figure 12.8).

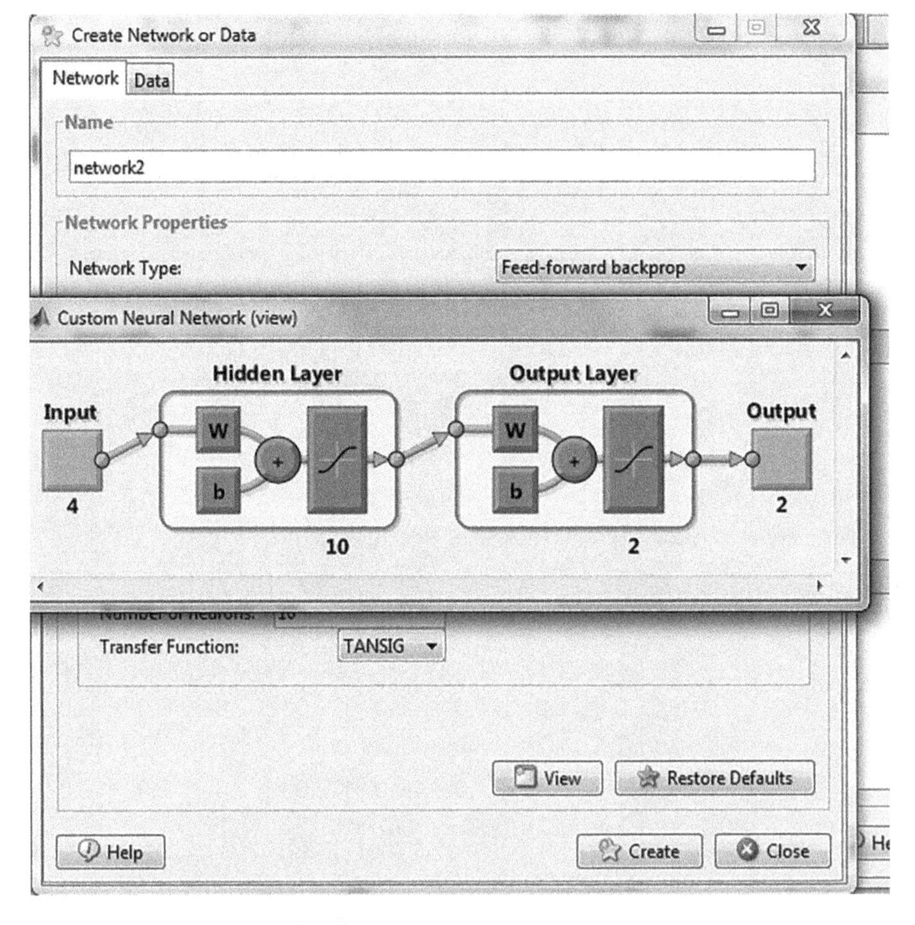

FIGURE 12.8
Generated neural network.

Now, the neural network can be seen in the Neural Network/Data Manager window, as shown in Figure 12.9.

Click 'network1' that is created, where we can see 'Train', 'Simulate' and other options. This is shown in Figure 12.10.

Step 3: Train the network and validate the test result

Click the 'Train' button and fill 'Training Info', i.e., input and target data and train the network (Figures 12.11 and 12.12).

Network Training till convergence and test result validation

In this stage, train the network until RC is close to 1 (RC = 1, ideal). We try to maximize the epoch and minimize the MSE (here, we take 'MSE' as a performance function). In this problem, we find the epoch at 12. Training performance plot is generated by clicking the 'Performance' button and regression plot is generated by clicking the 'Regression' button. These plots are shown in Figures 12.13 and 12.14.

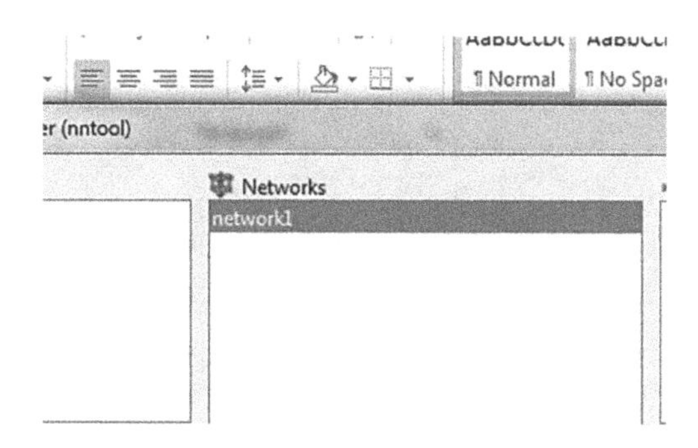

FIGURE 12.9
Neural Network/Data Manager window.

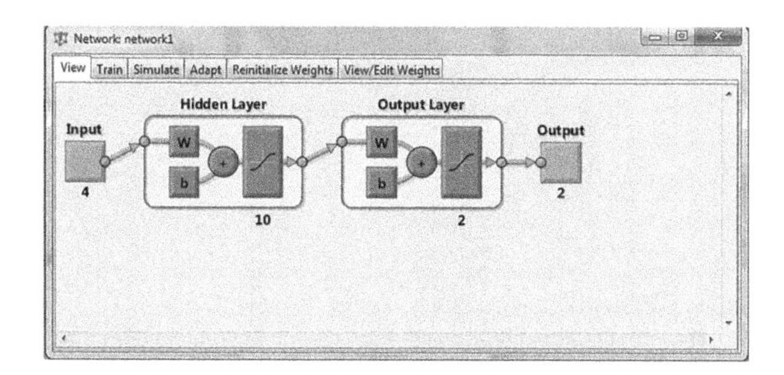

FIGURE 12.10
'Train' window.

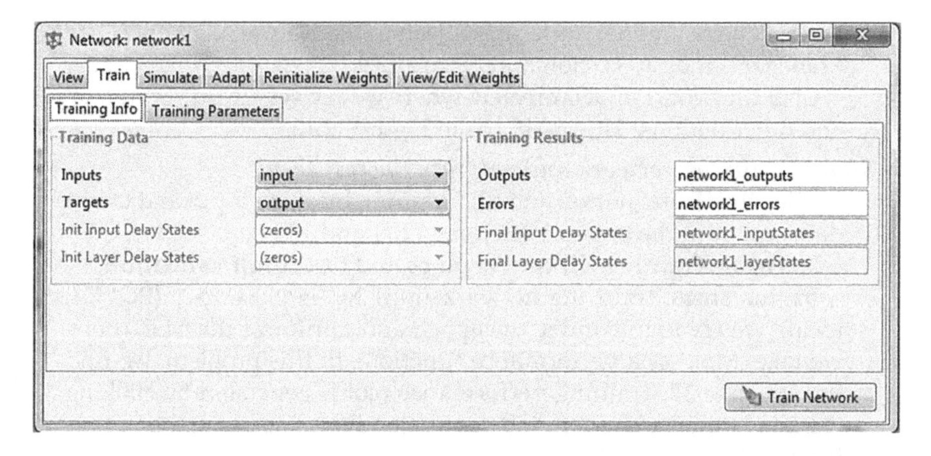

FIGURE 12.11
'Train' information input window.

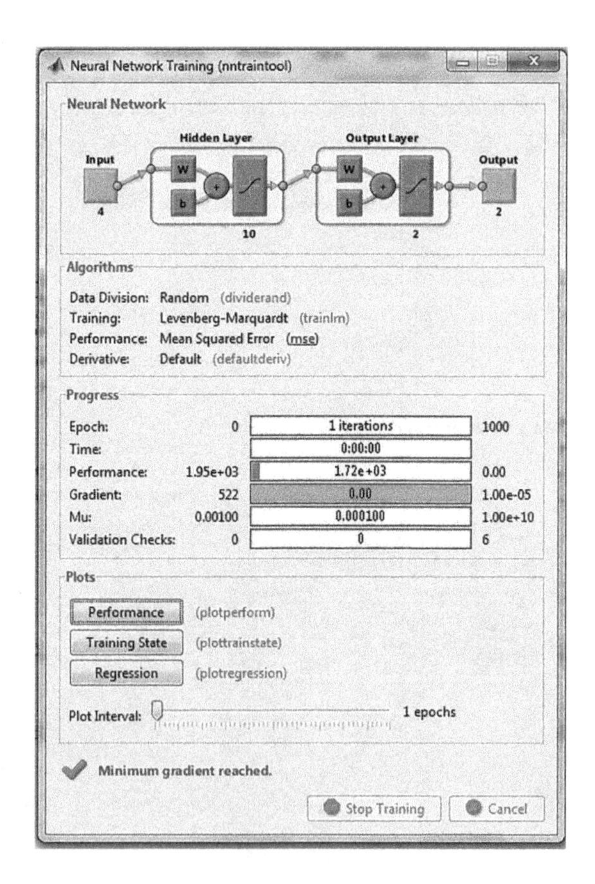

FIGURE 12.12
'Train' on-going process.

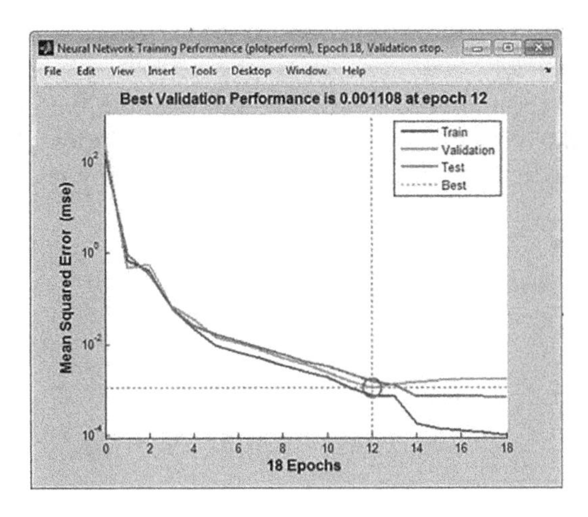

FIGURE 12.13
Generated output.

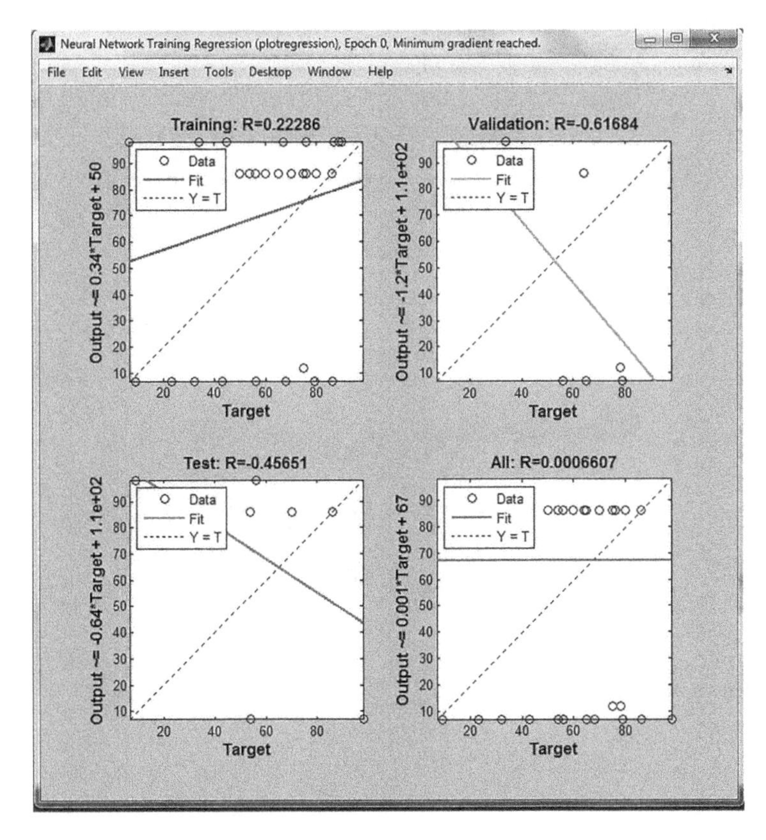

FIGURE 12.14
Various 'R' curves.

Step 4: Simulate the train network for the desired input and find the predicted value.

Simulate the network (Figure 12.15). The simulation results are stored in the Neural Network/Data Manager window (Figure 12.16).

It can also be imported to the workspace, as shown in Figure 12.17.

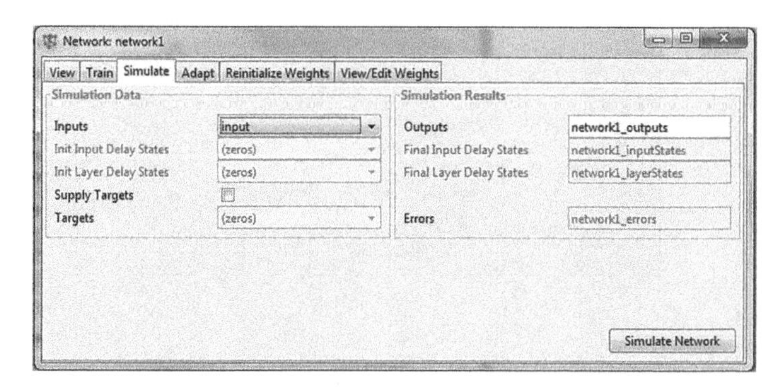

FIGURE 12.15
Simulation input and results.

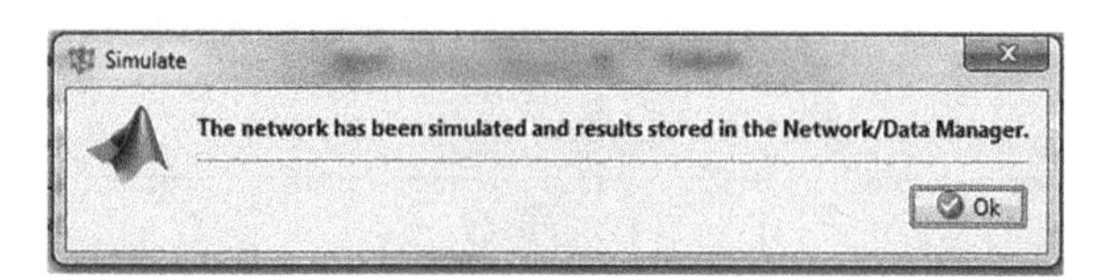

FIGURE 12.16
Window showing the data storage.

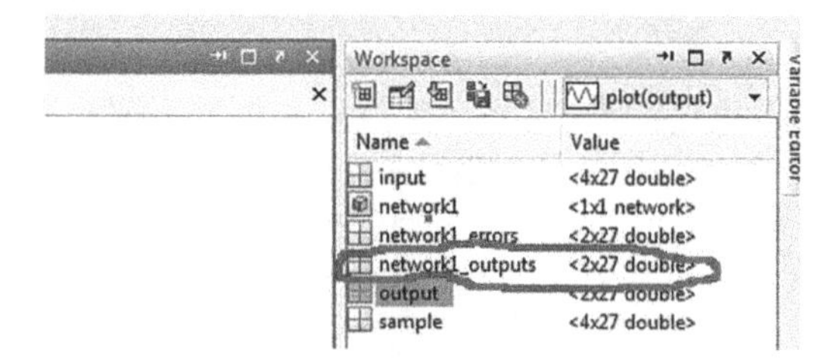

FIGURE 12.17
Retrieving output.

13

Working with MATLAB® for Optimization Using Genetic Algorithm

13.1 Introduction

Hands-on training on MATLAB®, a statistical software, is used to perform multi-objective optimization using Genetic Algorithm for a given problem.

This part is explained taking an example of two objective functions with two variables, as genetic algorithm (GA) is quite similar to multi-objective GA (*gamultiobj*). However, some differences also exist, which should be carefully understood before working with this tutorial.

Optimization using traditional GA is more general in nature. MATLAB utilizes *gamultiobj*, which is more specific and is slightly different from GA:

- *gamultiobj* uses Distance Measure Function, a function that indicates distance measurement for each individual with reference to its surrounding individuals.
- It undertakes a ratio called Pareto Fraction. It is between 0 and 1 and specifies the fraction of the density that is ensured while optimization.
- It takes only the *'Tournament selection function'*.
- It segregates non-inferior entities from the inferior ones and uses them. The better ones are automatically selected.
- It does not have a time constraint in achieving convergence.
- It contains various plot functions.
- It has a single scaling function.

13.2 Problem Definition

Multi-objective optimization with two outputs (*gamultiobj*)

Two functions, namely $f(1)$ and $f(2)$, are considered in this example, where the independent variable x has two values, namely x_1 and x_2.

13.3 Script Coding

Script code:

```
function f=mymulti1(x)
```
$f(1)=x_1^3-5*x_1^2+7*x_1*x_2-x_1^2*x_2^2+4x_2^3;$

$f(2)=x_1^3-10*x_1^2*x_2^2+x_1*x_2+7*x_2^3;$
```
end;
```

This script is required to be generated before operating on MATLAB.

13.4 Performing Optimization in MATLAB

Write the problem (writhing fitness function) in SCRIPT FILE and call the function as shown inFigure 13.1 (here *mymulti1* is the fitness function).

1. Options and conditions for the problem are stated as provided in Figures 13.2–13.4.
2. Optimization is initiated by using the Start command. Consequently, a graph (Pareto Front) appears as shown in Figure 13.5.

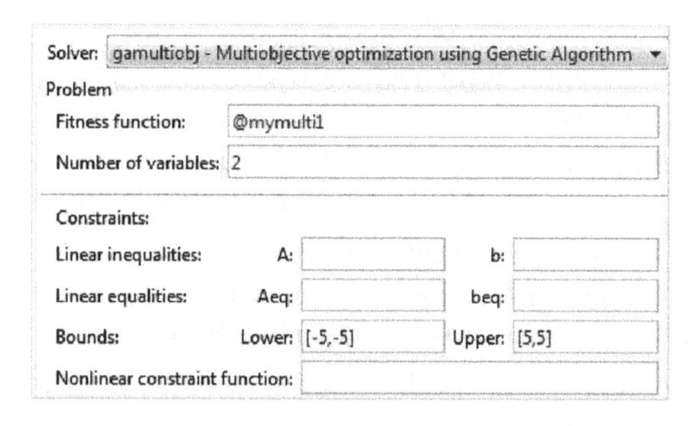

FIGURE 13.1
Initial window.

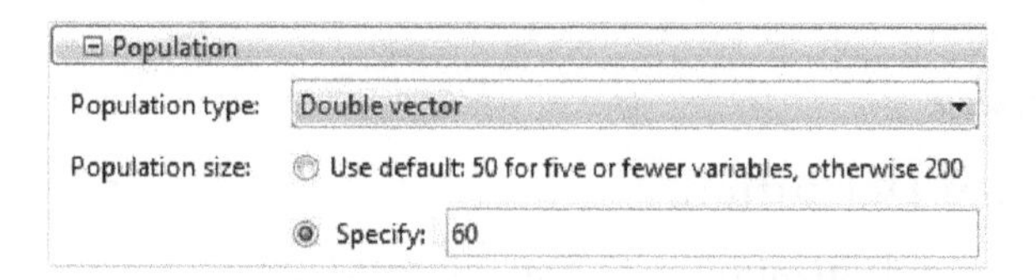

FIGURE 13.2
Population input.

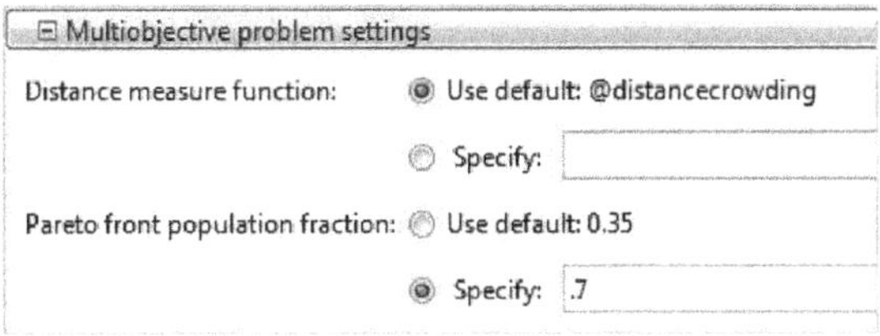

⊟ Multiobjective problem settings

Distance measure function: (•) Use default: @distancecrowding

( ) Specify:

Pareto front population fraction: ( ) Use default: 0.35

(•) Specify: .7

FIGURE 13.3
Multi-objective problem setting.

⊟ Plot functions

Plot interval: 1

[ ] Distance [ ] Genealogy [ ] Score diversity

[ ] Selection [ ] Stopping [✓] Pareto front

[ ] Average Pareto distance [ ] Rank histogram [ ] Average Pareto spread

[ ] Custom function:

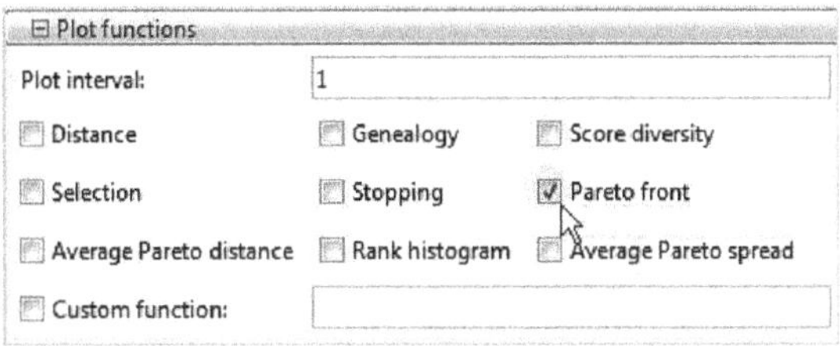

FIGURE 13.4
Plot function-choosing window.

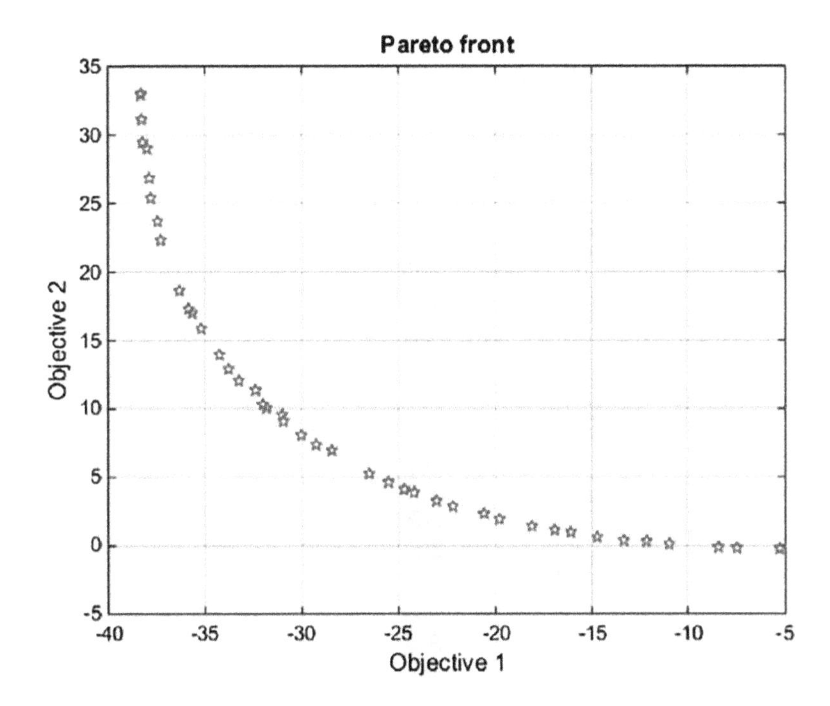

FIGURE 13.5
Pareto front.

Optimization running.
Optimization terminated: average change in the spread of Pareto solutions less than options.TolFun.

Pareto front - function values and decision variables

Index	f1	f2	x1	x2
1	-38.333	33.072	2.672	-1.976
2	-37.892	26.878	2.545	-1.802
3	-5.255	-0.25	0.707	-0.707
4	-32.333	11.368	2.09	-1.425
5	-7.526	-0.208	0.855	-0.795
6	-38.296	31.169	2.636	-1.977
7	-38.006	29.065	2.59	-1.808
8	-33.197	12.024	2.127	-1.6
9	-35.638	17.004	2.294	-1.79
10	-23.025	3.226	1.62	-1.335
11	-37.817	25.407	2.514	-1.904
12	-38.333	32.967	2.67	-1.976
13	-37.297	22.336	2.442	-1.825
14	-37.466	23.692	2.473	-1.764
15	-20.6	2.295	1.513	-1.322
16	-28.426	6.962	1.881	-1.585
17	-35.178	15.859	2.259	-1.761
18	-16.099	0.919	1.305	-1.174

FIGURE 13.6
Results output.

Pareto front - function values and decision variables

Index	f1	f2	x1	x2
1	-38.333	33.072	2.672	-1.976
12	-38.333	32.967	2.67	-1.976
6	-38.296	31.169	2.636	-1.977
28	-38.232	29.49	2.602	-1.937
7	-38.006	29.065	2.59	-1.808
2	-37.892	26.878	2.545	-1.802
11	-37.817	25.407	2.514	-1.904
14	-37.466	23.692	2.473	-1.764
13	-37.297	22.336	2.442	-1.825
27	-36.304	18.639	2.344	-1.765
41	-35.835	17.3	2.305	-1.73
9	-35.638	17.004	2.294	-1.79
17	35.178	15.859	2.259	1.761
21	-34.212	13.932	2.194	-1.73
29	-33.737	12.9	2.16	-1.645
8	-33.197	12.024	2.127	-1.6
4	-32.333	11.368	2.09	-1.425
23	-31.962	10.295	2.056	-1.537

Pareto front - function values and decision variables

Index	f1	f2	x1	x2
3	-5.255	-0.25	0.707	-0.707
5	-7.526	-0.208	0.855	-0.795
39	-8.466	-0.159	0.911	-0.79
20	-10.979	0.069	1.051	-0.818
26	-12.184	0.29	1.117	-1.106
31	-13.354	0.335	1.17	-1.002
30	-14.735	0.577	1.237	-1.066
18	-16.099	0.919	1.305	-1.174
32	-16.931	1.1	1.343	-1.005
19	-18.118	1.362	1.396	-1.146
37	-19.798	1.884	1.472	-1.194
15	-20.6	2.295	1.513	-1.322
36	-22.173	2.821	1.581	-1.205
10	-23.025	3.226	1.62	-1.335
42	-24.183	3.834	1.674	-1.383
38	-24.691	4.078	1.696	-1.373
40	-25.505	4.591	1.735	-1.42
22	-26.498	5.22	1.781	-1.441

FIGURE 13.7
Sorted results.

This graph indicates the difference between the two functions *f*1 and *f*2. It is plotted with respect to the objective function. The final optimized value is provided in the form of a table, shown in Figure 13.6, with various values of the input as well as the output, i.e., value of the functions.

Just like various common software packages, the values can be sorted in ascending or descending order by clicking any heading. Figure 13.7 shows the descending results of *f*1.

References

Asoh, H. & Muhlenbein, H., (1994). On mean convergence time of evolutionary algorithms without selection & mutation, in: *Parallel Problem Solving from Nature III*, Y. Davidor, H. P. Schwefel, & R. Männer, eds. Lecture Notes in Computer Science, Springer, Heidelberg, *866*: 98–107.

Back, T., (1995). Generalized convergence models for tournament-& (μ, λ)-selection, in: *Proceedings of the 6th International Conference on Genetic Algorithms*, Morgan Kaufmann, San Fransisco, CA, 2–8.

Back, T., Fogel, D. B. & Michalewicz, Z., (1997). *Handbook of Evolutionary Computation*, Oxford University Press, Oxford, UK.

Baker, J. E., (1985). Adaptive selection methods for genetic algorithms, in: *Proceedings of the International Conference on Genetic Algorithms & Their Applications*, Carnegie-Mellon University, Pittsburg, PA, 101–111.

Baluja, S., (1994). Population-based incremental learning: A method of integrating genetic search based function optimization & competitive learning, *Technical Report CMU-CS-94–163*, Carnegie Mellon University, USA.

Barthelemy, J. F. M. & Haftka, R. T., (1993). Approximation concepts for optimum structural design-a review, *Struct. Optim.*, *5*: 129–144.

Beasley, D., Bull, D. R. & Martin, R. R., (1993). An overview of genetic algorithms: fundamentals, *Univ. Comput.*, *15*: 58–69.

Blickle, T. & Thiele, L., (1995). A mathematical analysis of tournament selection, in: *Proceedings of the 6th International Conference on Genetic Algorithms*, Morgan Kaufmann, San Fransisco, CA, 9–16.

Booker, L. B., Fogel, D. B., Whitley, D. & Angeline, P. J., (1997). Recombination, in: *Handbook of Evolutionary Computation*, T. Black, D. B. Fogel, & Z. Michalewicz, eds., Chap. E3.3, IOP Publishing, Philadelphia, PA; Oxford University Press, Oxford, C3.3:1–C3.3:27.

Bosman, P. A. N. & Thierens, D., (1999). Linkage information processing in distribution estimation algorithms, in: *Proceedings of the 1999 Genetic & Evolutionary Computation Conference*, Orlando, Florida, USA, 60–67.

Bremermann, H. J., (1958) Evolution of intelligence. Nervous system as a model of its environment, *Technical Report No. 1, Department of Mathematics*, University of Washington, Seattle, WA.

Bulmer, M. G., (1985). *Mathematical Theory of Quantitative Genetics*, Oxford University Press, Oxford, UK.

Burke, E. K., Cowling, P. I., De Causmaecker, P. & VandenBerghe, G., (2001). A memetic approach to nurse roistering problem, *Appl. Intell.*, *15*: 199–214.

Burke, E. K., Elliman, D. G. & Weare, R. F., (1995). Specialized recombinative operators for timetabling problems, in: *Evolutionary Computing: AISB Workshop 1995*, T. Fogarty, ed., Lecture Notes in Computer Science, Springer, Berlin, *993*: 75–85.

Burke, E. K. & Newall, J. P., (1999). A multi-stage evolutionary algorithm for timetable problem, *IEEE Trans. Evol. Comput.*, *3*: 63–74.

Burke, E. K., Newall, J. P. & Weare, R. F. (1998). Initialization strategies & diversity in evolutionary timetabling, *Evol. Comput. (special issue on Scheduling)*, *6*: 81–103.

Burke, E. K. & Smith, A. J., (2000). Hybrid evolutionary techniques for maintenance scheduling problem, *IEEE Trans. Power Syst., 15*: 122–128.
Cantú-Paz, E., (1999). Migration policies & takeover times in parallel genetic algorithms, in: *Proceedings of the Genetic & Evolutionary Computation Conference,* Morgan Kaufmann, San Francisco, CA, 775–780.
Cantú-Paz, E., (2000). *Efficient & Accurate Parallel Genetic Algorithms,* Kluwer, Boston, MA.
Cheng, R. W. & Gen, M., (1997). Parallel machine scheduling problems using memetic algorithms, *Comput. Indust. Eng., 33*: 761–764.
Coley, D. A., (1999). *An Introduction to Genetic Algorithms for Scientists & Engineers,* World Scientific, New York.
Costa, D., (1995). An evolutionary tabu search algorithm in scheduling problem, *INFOR, 33*: 161–178.
Crow, J. F. & Kimura, M., (1970). *An Introduction of Population Genetics Theory,* Harper & Row, New York.
Davis, L., (1985). Applying algorithms to epistemic domains, in: *Proceedings of the International Joint Conference on Artificial Intelligence,* Los Angeles, California, USA, 162–164.
Davis, L. (ed.), (1987). *Genetic Algorithms & Simulated Annealing,* Pitman, London.
Davis, L. (ed.), (1991). *Handbook of Genetic Algorithms,* Van Nostrand Reinhold, New York.
Deb, K. & Goldberg, D. E., (1994). Sufficient conditions for deceptive & easy binary functions, *Artif. Intell., 10*: 385–408.
Falkenauer, E., (1998). *Genetic Algorithms & Grouping Problems,* Wiley, New York.
Fitzpatrick, J. M., Grefenstette, J. J. & VanGucht, D., (1984). Image registration by genetic search, in: *Proceedings of the IEEE Southeast Conference,* IEEE, Piscataway, NJ, 460–464.
Fleurent, C. & Ferland, J., (1994). Genetic hybrids for quadratic assignment problem, in: *DIMACS Series in Mathematics & Theoretical Computer Science, 16*: 190–206.
Fogel, D. B., (1998). *Evolutionary Computation: Fossil Record,* IEEE, Piscataway, NJ.
Forrest, S., (1993). Genetic algorithms: Principles of natural selection applied to computation, *Science, 261*: 872–878.
Fraser, A. S., (1957). Simulation of genetic systems by automatic digital computers. II: Effects of linkage on rates under selection, *Austral. J. Biol. Sci., 10*: 492–499.
Goldberg, D. E., (1987). Simple genetic algorithms & minimal deceptive problem, in: *Genetic Algorithms & Simulated Annealing,* L. Davis, ed., Chap. 6, Morgan Kaufmann, Los Altos, CA, 74–88.
Goldberg, D. E., (1989a). Genetic algorithms & Walsh functions: Part II, deception & its analysis, *Complex Syst., 3*: 153–171.
Goldberg, D. E., (1989b). *Genetic Algorithms in Search Optimization & Machine Learning,* Addison-Wesley, Reading, MA.
Goldberg, D. E., (1989c). Sizing populations for serial & parallel genetic algorithms, in: *Proceedings of the 3rd International Conference on Genetic Algorithms,* George Mason University, Fairfax, Virginia, USA, 70–79.
Goldberg, D. E., (1999a). Race, hurdle, & sweet spot: Lessons from genetic algorithms for automation of design innovation & creativity, in: *Evolutionary Design by Computers,* P. Bentley, ed., Chap. 4, Morgan Kaufmann, San Mateo, CA, 105–118.

Goldberg, D. E., (1999b). Using time efficiently: Genetic-evolutionary algorithms & continuation problem, in: *Proceedings of the Genetic & Evolutionary Computation Conference*, Orlando, Florida, USA, 212–219.

Goldberg, D. E., (2002). *Design of Innovation: Lessons from & For Competent Genetic Algorithms*, Kluwer, Boston, MA.

Goldberg, D. E. & Deb, K., (1991). A comparative analysis of selection schemes used in genetic algorithms, in: *Foundations of Genetic Algorithms*, G. J. E. Rawlins, ed., Morgan Kaufmann Publishers, San Mateo, CA, 69–93.

Goldberg, D. E., Deb, K. & Clark, J. H., (1992a). Genetic algorithms, noise, & sizing of populations, *Complex Syst.*, *6*: 333–362.

Goldberg, D. E., Deb, K. & Horn, J., (1992b). Massive multimodality, deception, & genetic algorithms, in: *Parallel Problem Solving from Nature II*, R. Männer & B. Manderick, eds., *Elsevier*, New York, 37–46.

Goldberg, D. E., Deb, K., Kargupta, H. & Harik, G., (1993). Rapid, accurate optimization of difficult problems using fast messy genetic algorithms, in: *Proceedings of the International Conference on Genetic Algorithms*, Urbana-Champaign, IL, USA, 56–64.

Goldberg, D. E., Korb, B. & Deb, K., (1989). Messy genetic algorithms: Motivation, analysis, & first results. *Complex Syst.*, *3*: 493–530.

Goldberg, D. E. & Lingle, R., (1985). Alleles, loci, & TSP, in: *Proceedings of the 1st International Conference on Genetic Algorithms*, Pittsburgh, PA, USA, 154–159.

Goldberg, D. E. & Rudnick, M., (1991). Genetic algorithms & variance of fitness, *Complex Syst.*, *5*: 265–278.

Goldberg, D. E. & Sastry, K., (2001). A practical schema theorem for genetic algorithm design & tuning, in: *Proceedings of the Genetic & Evolutionary Computation Conference*, San Fransisco, CA, 328–335.

Goldberg, D. E., Sastry, K. & Latoza, T., (2001). On supply of building blocks, in: *Proceedings of the Genetic & Evolutionary Computation Conference*, San Fransisco, CA, 336–342.

Goldberg, D. E. & Segrest, P., (1987). Finite Markov chain analysis of genetic algorithms, in: *Proceedings of the 2nd International Conference on Genetic Algorithms*, Cambridge, MA, USA, 1–8.

Goldberg, D. E. & Voessner, S., (1999). Optimizing global-local search hybrids, in: *Proceedings of the Genetic & Evolutionary Computation Conference*, Orlando, Florida, USA, 220–228.

Gorges-Schleuter, M., (1989). ASPARAGOS: An asynchronous parallel genetic optimization strategy, in: *Proceedings of the 3rd International Conference on Genetic Algorithms*, George Mason University, Fairfax, Virginia, USA, 422–428.

Gorges-Schleuter, M., (1997). ASPARAGOS96 & traveling salesman problem, in: *Proceedings of the IEEE International Conference on Evolutionary Computation*, Indianapolis, IN, USA, 171–174.

Grefenstette, J. J., (1981). Parallel adaptive algorithms for function optimization, Technical Report No. CS-81-19, *ComputerScience Department*, Vanderbilt University, Nashville.

Grefenstette, J. J. & Baker, J. E., (1989). How genetic algorithms work: A critical look at implicit parallelism, in: *Proceedingsof the 3rd International Conference on Genetic Algorithms*, George Mason University, Fairfax, Virginia, USA, 20–27.

Grefenstette, J. J. & Fitzpatrick, J. M., (1985). Genetic search with approximate function evaluations, in: *Proceedings of the International Conference on Genetic Algorithms & Their Applications*, Carnegie-Mellon University, Pittsburg, PA, 112–120.

Harik, G., Can't'u-Paz, E., Goldberg, D. E. & Miller, B. L., (1999). The gambler's ruin problem, genetic algorithms, & sizing of populations, *Evol. Comput., 7*: 231–253.

Harik, G., Lobo, F. & Goldberg, D. E., (1998). The compact genetic algorithm, in: *Proceedings of the IEEE International Conference on Evolutionary Computation*, Anchorage, Alaska, USA, 523–528.

Hart, W. E. & Belew, R. K., (1996). Optimization with genetic algorithm hybrids using local search, in: *Adaptive Individuals in Evolving Populations*, R. K. Belew & M. Mitchell, eds., Addison-Wesley, Reading, MA, 483–494.

Hart, W., Krasnogor, N. & Smith, J. E. (eds.), (2004). Special issue on memetic algorithms, *Evol. Comput., 12*(3): 345–352.

Heckendorn, R. B. & Wright, A. H., (2004). Efficient linkage discovery by limited probing, *Evol. Comput., 12*: 517–545.

Holland, J. H., (1975). *Adaptation in Natural & Artificial Systems*, University of Michigan Press, Ann Arbor, MI.

Ibaraki, T., (1997). Combinations with other optimization methods, in: *Handbook of Evolutionary Computation*, T. Bäck, D. B. Fogel, & Z. Michalewicz, eds., IOP Physics Publishing, Bristol; Oxford University Press, New York, D3:1–D3:2.

Kargupta, H., (1996). The gene expression messy genetic algorithm, in: *Proceedings of the International Conference on Evolutionary Computation*, Nayoya University, Japan, 814–819.

Krasnogor, N., Hart, W. & Smith, J. (eds.), (2004). *Recent Advances in Memetic Algorithms*, Studies in Fuzziness & Soft Computing, Springer, Berlin, *166*: 520–530.

Krasnogor, N. & Smith, J. E., (2005). A tutorial for competent memetic algorithms: Model, taxonomy & design issues, *IEEE Trans. Evol. Comput., 9*(5): 474–488.

Larrañaga, P. & Lozano, J. A. (eds.), (2002). *Estimation of Distribution Algorithms*, Kluwer, Boston, MA.

Lin, S. C., Goodman, E. D. & Punch, W. F., (1997). Investigating parallel genetic algorithms on job shop scheduling problem, in: *6th International Conference on Evolutionary Programming*, Indianapolis, IN, USA, 383–393.

Louis, S. J. & McDonnell, J., (2004). Learning with case injected genetic algorithms, IEEE Trans. *Evol. Comput., 8*: 316–328.

Man, K. F., Tang, K. S. & Kwong, S., (1999). *Genetic Algorithms: Concepts & Design*, Springer, London.

Manderick, B. & Spiessens, P., (1989). Fine-grained parallel genetic algorithms, in: *Proceedings of the 3rd International Conference on Genetic Algorithms*, George Mason University, Fairfax, Virginia, USA, 428–433.

Michalewicz, Z., (1996). *Genetic Algorithms + Data Structures = Evolution Programs*, 3rd edn., Springer, Berlin.

Miller, B. L. & Goldberg, D. E., (1995). Genetic algorithms, tournament selection, & effects of noise, *Complex Syst., 9*: 193–212.

Miller, B. L. & Goldberg, D. E., (1996a). Genetic algorithms, selection schemes, & varying effects of noise, *Evol. Comput., 4*: 113–131.

Miller, B. L. & Goldberg, D. E., (1996b). Optimal sampling for genetic algorithms, *Intelligent Engineering Systems through Artificial Neural Networks (ANNIE'96)*, ASME Press, New York, *6*: 291–297.

Mitchell, M., (1996). *Introduction to Genetic Algorithms*, MIT Press, Boston, MA.

Mluhlenbein, H. & Paaß, G., (1996). From recombination of genes to estimation of distributions I. Binary parameters, in: *Parallel Problem Solving from Nature IV*, H. M. Voigt, W. Ebeling, I. Rechenberg, & H. P. Schwefel eds., Lecture Notes in Computer Science, Springer, Berlin, *1141*: 178–187.

Mluhlenbein, H. & Schlierkamp-Voosen, D., (1993). Predictive models for breeder genetic algorithm, *Evol. Comput., 1*: 25–49.

Moscato, P., (1989). On evolution, search, optimization, genetic algorithms & martial arts: Towards memetic algorithms, *Technical Report C3P 826, Caltech Concurrent Computation Program*, California Institute of Technology, Pasadena, CA.

Moscato, P., (1999). Memetic algorithms, in: *New Ideas in Optimization*, D. Corne, M. Dorigo, & F. Glover, eds., McGraw-Hill, New York.

Moscato, P., (2001). Memetic algorithms, in: *Handbook of Applied Optimization*, Section 3.6.4, P. M. Pardalos & M. G. C. Resende, eds., Oxford University Press, Oxford.

Moscato, P. & Cotta, C., (2003). A gentle introduction to memetic algorithms, in: *Handbook of Metaheuristics*, F. Glover & G. Kochenberger, eds., Chap. 5, Kluwer, Norwell, MA.

Munetomo, M. & Goldberg, D. E., (1999). Linkage identification by nonmonotonicity detection for overlapping functions, *Evol. Comput., 7*: 377–398.

Oliver, J. M., Smith, D. J. & Holland, J. R. C., (1987). A study of permutation crossover operators on travelling salesman problem, in: *Proceedings of the 2nd International Conference on Genetic Algorithms*, Cambridge, MA, USA, 224–230.

Paechter, B., Cumming, A. & Luchian, H., (1995). The use of local search suggestion lists for improving solution of timetable problems with evolutionary algorithms, in: *Evolutionary Computing: AISB Workshop 1995*, T. Fogarty, ed., Lecture Notes in Computer Science, Springer, Berlin, *993*: 86–93.

Paechter, B., Cumming, A., Norman, M. G. & Luchian, H., (1996). Extensions to a memetic timetabling system, in: *Practice & Theory of Automated Timetabling I*, E. K. Burke & P. Ross, eds., Lecture Notes in Computer Science, Springer, Berlin, *1153*: 251–265.

Pelikan, M., (2005). *Hierarchical Bayesian Optimization Algorithm: Toward a New Generation of Evolutionary Algorithm*, Springer, Berlin.

Pelikan, M. & Goldberg, D. E., (2001). Escaping hierarchical traps with competent genetic algorithms, in: *Proceedingsof the Genetic & Evolutionary Computation Conference*, San Fransisco, CA, 511–518.

Pelikan, M., Goldberg, D. E. & Cantú-Paz, E., (2000). Linkage learning, estimation distribution, & Bayesian networks, *Evol. Comput., 8*: 314–341.

Pelikan, M., Lobo, F. & Goldberg, D. E., (2002). A survey of optimization by building & using probabilistic models, *Comput. Optim. Appl., 21*: 5–20.

Pelikan, M. & Sastry, K., (2004). Fitness inheritance in Bayesian optimization algorithm, in: *Proceedings of the Genetic & Evolutionary Computation Conference*, Seattle, Washington, USA, *2*: 48–59.

Pettey, C. C., Leuze, M. R. & Grefenstette, J. J., (1987). A parallel genetic algorithm, in: *Proceedings of the 2nd International Conference on Genetic Algorithms*, Cambridge, MA, USA, 155–161.

Radcliffe, N. J. & Surry, P. D., (1994). Formal memetic algorithms, in: *Evolutionary Computing: AISB Workshop 1994*, T. Fogarty, ed., Lecture Notes in Computer Science, Springer, Berlin, *865*: 1–16.

Reeves, C. R., (1995). Genetic algorithms, in: *Modern Heuristic Techniques for Combinatorial Problems*, C. R. Reeves, ed., McGraw-Hill, New York.

Robertson, G. G., (1987). Parallel implementation of genetic algorithms in a classifier system, in: *Proceedings of the 2nd International Conference on Genetic Algorithms*, Cambridge, MA, USA, 140–147.

Rothlauf, F., (2002). *Representations for Genetic & Evolutionary Algorithms*, Springer-Verlag, Berlin.

Rudolph, G., (2000). Takeover times & probabilities of non-generational selection rules, in: *Proceedings of the Genetic & Evolutionary Computation Conference*, Las Vegas, Nevada, USA, 903–910.

Sakamoto, Y. & Goldberg, D. E., (1997). Takeover time in a noisy environment, in: *Proceedings of the 7th International Conference on Genetic Algorithms*, East Lansing, MI, USA, 160–165.

Sastry, K. & Goldberg, D. E., (2002). Analysis of mixing in genetic algorithms: A survey, *IlliGAL Report No. 2002012*, University of Illinois at Urbana Champaign, Urbana, IL.

Sastry, K. & Goldberg, D. E., (2003). Scalability of selector combinative genetic algorithms for problems with tight linkage, in: *Proceedings of the 2003 Genetic & Evolutionary Computation Conference*, Chicago, Illinois, USA, 1332–1344.

Sastry, K. & Goldberg, D. E., (2004a). Designing competent mutation operators via probabilistic model building of neighborhoods, in: *Proceedings of the 2004 Genetic & Evolutionary Computation Conference II*, K. Deb, ed., Lecture Notes in Computer Science, Springer, Berlin, *3103*: 114–125.

Sastry, K. & Goldberg, D. E., (2004b). Let's get ready to rumble: Crossover versus mutation head to head, in: *Proceedings of the 2004 Genetic & Evolutionary Computation Conference II*, K. Deb, ed., Lecture Notes in Computer Science, Springer, Berlin, *3103*: 126–137.

Sastry, K., Goldberg, D. E., & Pelikan, M., (2001). Don't evaluate, inherit, in: *Proceedings of the Genetic & Evolutionary Computation Conference*, San Fransisco, CA, 551–558.

Sastry, K., Pelikan, M. & Goldberg, D. E., (2004). Efficiency enhancement of genetic algorithms building-block-wise fitness estimation, in: *Proceedings of the IEEE International Congress on Evolutionary Computation*, Portland, OR, USA, 720–727.

Smith, R., Dike, B. & Stegmann, S., (1995). Fitness inheritance in genetic algorithms, in: *Proceedings of the ACM Symposium on Applied Computing*, New York, 345–350.

Spears, W., (1997). Recombination parameters, in: *Handbook of Evolutionary Computation*, T. Black, D. B. Fogel, & Z. Michalewicz, eds., Chap. E1.3, IOP Publishing, Philadelphia, PA; Oxford University Press, Oxford, E1.3:1–E1.3:13.

Spears, W. M. & De Jong, K. A., (1994). On virtues of parameterized uniform crossover, in: *Proceedings of the 4th International Conference on Genetic Algorithms*, Urbana-Champaign, USA.

Srivastava, R. & Goldberg, D. E., (2001). Verification of theory of genetic & evolutionary continuation, in: *Proceedings of the Genetic & Evolutionary Computation Conference*, San Fransisco, CA, 551–558.

Syswerda, G., (1989). Uniform crossover in genetic algorithms, in: *Proceedings of the 3rd International Conference on Genetic Algorithms*, George Mason University, Fairfax, Virginia, USA, 2–9.

Thierens, D., 1999, Scalability problems of simple genetic algorithms, *Evol. Comput.*, *7*: 331–352.

Thierens, D. & Goldberg, D. E., (1994a). Convergence models of genetic algorithm selection schemes, in: *Parallel Problem Solving from Nature III*, 116–121.

Thierens, D. & Goldberg, D. E., (1994b). Elitist recombination: An integrated selection recombination GA, in: *Proceedings of the 1st IEEE Conference on Evolutionary Computation*, Orlando, Florida, USA, 508–512.

Thierens, D., Goldberg, D. E. & Pereira, A. G., (1998). Domino convergence, drift, & temporal-salience structure of problems, in: *Proceedings of the IEEE International Conference on Evolutionary Computation*, Anchorage, Alaska, USA, 535–540.

Valenzuala, J. & Smith, A. E., (2002). A seeded memetic algorithm for large unit commitment problems, *J. Heuristics, 8*: 173–196.

Voigt, H. M., Mluhlenbein, H. & Schlierkamp-Voosen, D., (1996). The responses to selection equation for skew fitness distributions, in: *Proceedings of the International Conference on Evolutionary Computation*, Nayoya University, Japan, 820–825.

Watson, J. P., Rana, S., Whitely, L. D. & Howe, A. E., (1999). The impact of approximate evaluation on performance of search algorithms for warehouse scheduling, *J. Scheduling,* 2: 79–98.

Whitley, D., (1995). Modeling hybrid genetic algorithms, in *Genetic Algorithms in Engineering & Computer Science*, G. Winter, J. Periaux, M. Galan, & P. Cuesta, eds., Wiley, New York, 191–201.

Yu, T. L., Goldberg, D. E., Yassine, A. & Chen, Y. P., (2003). A genetic algorithm design inspired by organizational theory: Pilot study of a dependency structure matrix driven genetic algorithm, *Artificial Neural Networks in Engineering (ANNIE 2003)*, St. Louis, Missouri, USA, 327–332.

Index